Love from the Other Side

CAROL SHIMP

Love from the Other Side

SEARCHING FOR ANSWERS

To Jodie
Blessings
Carol Shimp

Outskirts Press, Inc.
Denver, Colorado

The opinions expressed in this manuscript are solely the opinions of the author and do not represent the opinions or thoughts of the publisher. The author has represented and warranted full ownership and/or legal right to publish all the materials in this book.

Love From the Other Side
Searching for Answers
All Rights Reserved.
Copyright © 2011 Carol Shimp
v2.0 r1.0

Cover Photo © 2011 JupiterImages Corporation. All rights reserved - used with permission.

This book may not be reproduced, transmitted, or stored in whole or in part by any means, including graphic, electronic, or mechanical without the express written consent of the publisher except in the case of brief quotations embodied in critical articles and reviews.

Outskirts Press, Inc.
http://www.outskirtspress.com

ISBN: 978-1-4327-6790-7

Library of Congress Control Number: 2011921049

Outskirts Press and the "OP" logo are trademarks belonging to Outskirts Press, Inc.

PRINTED IN THE UNITED STATES OF AMERICA

Contents

Introduction vii
Chapter 1:   Spiritual Childhood 1
Chapter 2:   Many Years Later 11
Chapter 3:   Danny 17
Chapter 4:   Returning Home 23
Chapter 5:   The Phone Calls 25
Chapter 6:   Alex 37
Chapter 7:   Taking Walks 53
Chapter 8:   Seeking Help 57
Chapter 9:   The Haven for Spiritual Travelers 63
Chapter 10: Spirits 75
Chapter 11: The Supermarket 79
Chapter 12: Janet 83
Chapter 13: Automatic Writing 91
Chapter 14: Karma 97

Chapter 15: The Accident 105
Chapter 16: Where Angels Fear to Tread 111
Chapter 17: The Dream/Thanksgiving 117
Chapter 18: December and Christmas 125
Chapter 19: Henry 135
Chapter 20: January New Year's/Yard Sale 141
Chapter 21: Crossing Over 151
Chapter 22: Letting Go 157
Chapter 23: Publix 165
Chapter 24: The Gift 171
Chapter 25: Ohio 183
Chapter 26: Footprints 197
Chapter 27: Back to Work 203
Chapter 28: Thomas O'Leary 211
Chapter 29: Christmas 221
Chapter 30: Regression 225
Chapter 31: Love 241
Afterword 245
Acknowledgements 247
Books That Have Helped Me 249

Introduction

My name is Carol Shimp and this is my story. Some names have been changed to protect the privacy of family and characters. I have used fiction in portions and edited for clarity. My characters from Publix are fictional. This story is told according to my memory, my experiences, and notes I kept. I'm just an ordinary person having an extraordinary spiritual life.

My experience was with the earthbound entity I have, named Danny. He entered my energy field and brought back all of the memories I had repressed. We (I) worked with his spirit and we helped each other to find healing, peace, light, and love.

CHAPTER 1

Spiritual Childhood

A young child's viewpoint is innocent and non-judgmental. Anything and everything can be new and wonderful until years of conditioning start us questioning what we see and hear, coloring how we interpret life. My family life stayed a little more in tune with the open-minded view of existence. Things happened regularly in my family that kept me watching and open for conditions and events that don't quite fit in the accepted normal worldview.

In the late '40s, when I was six years old, televisions weren't in every home yet. Being a typical child of the time, I wanted to watch *Howdy Doody* like all my friends did, so I used to visit our neighbor, Mrs. Vesta Shall. Mr. and Mrs. Shall happened to have the closest TV as well as a generous nature and two sons who had been active military in World War II. One evening I stood by her dining room table and saw her son Teddy standing on the stairs. Being a friendly child I looked up and said, "Hi, Teddy!" Behind me, Mrs. Shall burst into sobs and tears. She called my mother and I was rushed out of the house. I was confused; I didn't know what I had done to cause Mrs. Shall to cry, and I have never remembered exactly what my mother said to me at the time. It wasn't until much later that I found out Teddy had died in the war.

Such things have occurred my entire life. I remember an afternoon in my middle forties when I was helping my mother clean her refrigerator at her home in South Florida. The side kitchen door was open. Suddenly a shadow passed through the side door, through me, and continued out the front door.

I asked my mother, who was nearby, "Did you see that?"

"See what?" she asked.

By then, I was used to the fact that not everyone could see or sense my spiritual experiences. I just figured whatever "It" was, it was trying to get somewhere. When I heard voices in my sleep or saw a vision, I knew it was a premonition of things to come and nothing to stress over. It was simply the way my life operated.

I was born on the 30th of June 1941, and grew up in Canton, Ohio. My mother said she thought I was a tumor. She didn't think she was pregnant because I was her change of life baby. My mother told me that when I was born my aunts couldn't get over my large liquid brown eyes. "Carol has the most beautiful eyes," my Aunt Ruth would say.

In the late 1940s and early 1950s, the American Midwest wasn't all one big Norman Rockwell painting, but there were places and times that fit the profile of his artwork -- especially when family would get together.

I can almost smell the purple lilacs from our yard that bloomed in early spring. Typical of the era, my father would get creative from time to time with his can-do attitude, making something out of nothing. Once, he scavenged a dark blue back seat he had removed from an old car, and built a swing with a wood frame around it. He then hung two chains from the beams in the ceiling of our front porch.

During the summers when I was six or seven years old, my mother would fix a light lunch, sometimes egg salad sandwiches and juicy peaches ripened to perfection, with glasses of refreshing iced tea with lemon. Afterwards, on the cool summer afternoons, I would stretch out on our porch swing with my eyes half closed and listen to nearby traffic until it would seem to drift far away. A gentle breeze would rock the swing, while I dreamed I held a golden string attached to a large white helium-filled balloon floating and lifting me upward toward the sky. That's where I would meet my love, my prince. Then the sharp barking of a dog or the harsh blare of an automobile horn would bring me back to earth. My balloon always burst just when I would get close to seeing my prince.

I wasn't the only one to enjoy that wonderful swing. On many late, lazy summer afternoons, I would find my mother or sisters stretched out there, rocked sound asleep by soft summer breezes. Our front porch was a gathering spot, the stage where stories were told. On cool summer evenings, the rest of the family gathered there. My aunts, uncles, and cousins would settle in with us on the porch, where we passed around large bowls of popcorn. We would sit, sipping iced tea and sodas, while we listened to the old family stories, along with a few current ones. They were often about old superstitions and spirits.

When I was six years old, I remember my mother told me, "Grandpa Joe [who lived with us in Ohio], is going home to Pennsylvania for a visit."

I was worried because he was leaving us. Immediately I started toward the back yard and found Grandpa Joe in his baggy black slacks and white-collared shirt, sitting on an old wooden bench under the apple tree.

I sat down beside him and carefully smoothed my blue dress. Mother was always telling me to keep my knees together and my skirt down. I watched as he carefully peeled a shiny red apple with his pocketknife. He cut a piece off and offered it to me. The taste was a little tart and I made a face. "Grandpa, since you came to live with us you have never gone back home to Pennsylvania. I don't want you to go."

He wiped a tear from my cheek. "I have to go home. It's time," he said with a smile.

"But I'll never see you again," I cried.

"Of course you will. If you need me, all you have to do is think of me." He carefully placed his pocketknife and apple on the bench beside him. Then with one hand he brushed a strand of gray hair off his forehead before reaching his arms out to hug me. I looked in his liquid brown eyes and wrapped my arms around his neck and held on real tight. I cried hard because although Mom told me he was going for a visit, somehow I knew, Grandpa Joe wasn't coming back to us. I don't know how I knew. But I knew. He left later by car with my Aunt Ruth.

When Mother answered the phone call, and was told the news, we packed a suitcase and climbed in my dad's Ford, then headed for Pennsylvania. My grandpa, Joseph Wisniewski, had a heart attack on the steps of the Polish Club he founded in New Kensington, Pennsylvania. I had cried so hard at his funeral I remember everyone kept handing me handkerchiefs for my nose and eyes. We buried him next to Grandma.

While growing up, I had plenty of time to daydream. My sister Vivian was twelve years older than I, and Emily had me by ten years. By the time I was old enough to play on my own, Vivian and Emily were busy with high school, and working weekends in the neighborhood soda shop.

I remember one afternoon in my preschool years something drew me inside my bedroom closet. I don't know how long I slept there, but I heard Mother call me. She found me on the floor in the corner.

"Carol, what are you doing in the closet? I have been looking all over the neighborhood for you." It was a few minutes before I realized where I was, still groggy from a deep sleep. "I had the neighbors searching for you."

I didn't understand why they were worried about me.

Often my mother used to whisper to my aunts or speak Polish when she didn't want me to hear her conversations. I sensed she was discussing something about family and spirits. I heard bits and parts of conversations. I believed that was why Father Burnis, a priest from All Saints Polish Catholic Church, would grace our family with a house blessing. My mother would empty and beat the dusty bag from her Hoover vacuum cleaner against the side of our home, a cloud of dust filling the air. She would be in high spirits, singing "O Come All Ye Faithful" as she swept and dusted our home. Mother always hummed Christmas carols when she cleaned our home. I believed those were the only songs she knew the words to. I knew someone special was coming to visit. Once a year around Easter, we would meet Father Burnis at our front door. He always wore black pants and shirt with his white collar, his crucifix on a chain around his neck. I watched the back of his head and his thinning white hair as he wrote on the top frame of our front door with white chalk: 20 C + M + B 09. Then he would recite a prayer, "Oh Blessed Trinity." After the blessing, Father Burnis would visit with us. My dad retrieved from our basement a bottle of his homemade red wine, and shared it with Father Burnis. Mother and I went about doing other things while my dad and our priest sat at our kitchen table and talked. Our priest had rosy cheeks and was feeling jolly when he was ready to leave.

"God bless and protect you, Marge and Frank." He waved goodbye and walked toward his car.

Attending first grade, I walked several blocks to Saint Paul's grade school. Sometimes I walked with friends and other times by myself. Someone, it seemed, was always with me…nobody I could see, but it was an energy and feeling I had. A nun in her long black dress and white habit told our catechism class about guardian angels that protected us. I accepted the fact that I had a guardian angel. In the second grade after my first communion, I walked to confession on Saturday afternoons. It was safe to walk in those days. One Saturday the nuns had locked the bathrooms. When I arrived in church I really had to relieve myself, but the restrooms were locked so I tried to hold it. I entered the confessional and blessed myself.

LOVE FROM THE OTHER SIDE

I began with "Bless me, Father, for I have sinned." Just as I said the words I felt warm liquid running down my legs. I ended up peeing in the confessional. The priest never said anything to me. After that the bathrooms were not locked. My underpants must have dried while walking home in the hot summer afternoon. I was too embarrassed to tell my mother. She never asked me about the incident.

At eight years of age, on a Saturday, Emily took me to a baseball game played on a field in our neighborhood. There was a gas station beside the field where the kids hung out. Two brothers owned the station, and they had a German Shepherd dog. I was standing in the station with the other children when suddenly the dog lunged toward me and ripped my face. My upper lip was hanging below my bottom one. I must have been terrified; with all the commotion, it's hard for me to remember. They rushed me to the hospital. I couldn't figure out what I did to make the dog angry. I have never been afraid of animals and I felt sorry for that dog because they put it to sleep. The doctor did reconstructive surgery on my face. I attended school while my face was still bandaged. I don't remember how my classmates reacted toward me. I felt my lay teacher showed me sympathy. Mother didn't want me to look in a mirror, but I did. Looking in the mirror, I wondered if I would ever look normal again.

Dear Guardian Angel, will I ever be pretty again? What will my friends think of me? With the back of my hand I wiped the tears from my face. As time went by my scars were less visible. Emily told me recently she always felt guilty because she took me to the ball game and was there at the station. It wasn't her fault. We have no control over incidents, they just happen -- but now when I hear a dog bark viciously, I freeze.

There were better times. We lived on Fifteenth Street, just off of Harrisburg Road. James Dry Cleaning and Laundry stood across the street from our home. Vivian and Emily used to look out of their bedroom window checking out the truck drivers as they came and left from the parking garage. I overheard their discussions, normal domestic subjects: "Oh, Joe's single," or "Tom is married." The owners of the business had a daughter almost my age, and we became friends. In the evening after all of the employees left for the day, we played. There was a concrete ramp leading from the dry-cleaning room to the laundry room. April and I took turns crawling in the canvas buggies. Then we pushed each other down the ramp. We would go flying down the aisle between the washers and curtain stretchers, the wheels roaring past the aroma of laundry soap and bleach. The following morning when employees came to work they had a fit, because the laundry orders were mixed up. After that, April's dad told us to use our roller skates. Some days, when we took the time to untangle a confused mass of hangers piled in a corner, we were paid a penny a hanger. We chatted and giggled as we worked.

Down the road from the laundry stood Quality Dairy where milk was processed. Some days I would stop there for chocolate milk. The far end of our street was a creek. Behind the creek was a factory, Union Metal. My cousins and I used to wade in the creek, our bare feet in water with black muck oozing between our toes. We would catch tadpoles in a jar and sometimes minnows. On weekends we would go the local Windsor movie theater to see Roy Rogers and Dale Evans, or Gene Autry.

My mother, Marge, spent her life taking care of others. When my grandmother passed away, my grandfather and Aunt Fay (who was twelve years old at the time) came to live with us. Not too long after that, my father's niece Paula came to our home. After my mother's sister died, Uncle Larry came to join us in our two-story stucco home as well.

My mother and father, Frank, worked in the steel mill during World War II. Our immediate and extended family never worried about dinner. Whether it was tuna and cream sauce, potato pancakes, or pot roast and potatoes, we all gathered around the dinner table for Mother's cooking. She was always baking or busy with pots on the stove. My sisters Vivian and Emily and I were usually stuck doing dishes.

◄ LOVE FROM THE OTHER SIDE

As I grew older I suppressed my strange feelings and premonitions. I dated guys in high school but none of them interested me. Then I met Danny Malone. We dated and he gave me an engagement ring in my senior year. I finished high school but wanted more out of life. I broke up with Danny and at eighteen years of age I packed my bags and left for New York City. After two years I returned home, was introduced to and married Alex Shimp. After that, marriage, children, and everyday life kept me busy.

While searching through boxes of keepsakes recently, I found a poem on the back page of Mother's tattered recipe book. The poem says a lot about my mother and her spirit.

THE KITCHEN PRAYER

Lord of all pots and pans, and things,
Since I've not time to be
A saint by doing lovely things
Or watching late with Thee,
Or dreaming in the dawn light,
Or storming Heaven's gates,
Make me a saint by getting meals
And washing up the plates
Although I must have Martha's hands
I have a Mary mind,
And when I black the boots and shoes,
Thy sandals, Lord, I find
I think of how they trod the earth
Each time I scrub the floor;
Accept this meditation, Lord -
I haven't time for more.

Warm all the kitchen with Thy love
And light it with Thy peace.
Forgive me all my worry,

And make my grumbling cease.
Thou who didst love to give men food
In room or by the sea,
Accept this service I do.
I do it unto Thee.

--Klara Munkres

Since moving to South Florida in 1969, most of my relatives have passed on. I miss those family gatherings we had on the front porch. My family taught me to be open-minded and accepting of the mysteries of spirit and life. So when I am walking on a warm day and a cool breeze ruffles my hair I feel blessed.

All my life, I have had dreams and messages, feelings that tended to be precognitive in nature. In 1978, after my mother Marge, passed away, her active spirit, and the love she sent me triggered my need for knowledge and experience of the other side. Although I meditated privately, I also found a small group who gathered in a circle to meditate. Together we brought in white light for love, peace, and healing. I read New Age books on spirit materialization and dreams. It's not really New Age. People have been giving prophecies for centuries. In the 1950s, during my girlhood, people didn't even talk about those things. But in 1996 on Mother's Day in May, my spirit guides came to remind me I needed to pay attention. I was in complete shock. Being raised a Catholic, I still went to church and prayed my rosary. I learned self-discipline in Catholic School. Mass and receiving Holy Communion always made me feel saintly. I never discussed my spirit friends with a priest. I wanted to keep peace in my two conflicting worlds. Why create a controversy I knew I couldn't win? I wasn't sure what my spirit guides were trying to tell me. But I knew I wanted to go on with my quest for learning. None of my studies had prepared me for the experience I was about to have.

CHAPTER 2

Many Years Later

APRIL

I had been living in southern Florida for twenty-seven years. It was early morning, Easter Sunday, 1996. I was in a deep sleep. In my dream, I was having an out of body experience, astral floating on a cloud. While I was there, God's warm bright yellow light surrounded me and filled me with pure love. I was given instructions. I couldn't remember all of the instructions, but I instinctively knew it was something I had to do. The message contained something about my past, present, and future. It was something I had to do to complete my soul growth. I heard my Spirit Guide say, *"He has the gift. It will be all right. He loves you."*

I heard myself say, *"I don't understand."*

Before I knew it, my astral body returned to my physical body. It felt like I was gliding down a waterslide and there I was, home again. I awoke wondering what I had to do. I wanted to remember, to understand. I lay in bed, reluctant to get up. All I could remember was "He has the gift. It will be all right. He loves you." It must have been my prophecy, but my dream left me feeling good. The rest of my message was gone.

May

As I look back my prophecy began on Sunday, Mother's Day. Little did I know how much I would learn, and what little knowledge I had. As events happened, it seemed like déjà vu.

It was a typical warm, sunny morning. I heard our German Shepherd, Princess, barking. I glanced out the front window and saw my thirty-two-year-old daughter Janet park her car in our driveway and honk the horn. Janet had scheduled a day off from work so we could go out and play.

"I'll be right out," I yelled through the open front window.

We were leaving our husbands Alex and Henry, and my son Jerry in the house for the day to do whatever guys do when they are left alone. I was happy to have a day out. I climbed into her 1985 maroon Camaro and we headed for the local Swap Shop flea market on Sunrise Boulevard. Janet told me she was going to buy my Mother's Day gift there. I glanced at my daughter as she pulled out from the curb.

"Did you check your blood sugar?" I asked Janet. "If we're going to shop all day, I want you to be well. You know I worry about you."

"Mother, I'm grown up now. I take good care of myself."

"Janet when it comes to parents, children will always be children. How can I not worry about you? I love you." We looked at each other and grinned. "Now that we've gotten that out of the way, let's have fun. I want a hairdini so I can put my hair up in a twist. Maybe I could also use some hair dye to cover my gray."

"Yes, a twist would be good."

I looked in the mirror behind the visor. Gray strands among the light brown. Not too bad, I thought for a fifty-plus woman. We both had our hair in ponytails. I fluffed mine as I watched her take a drag from her cigarette. I thought, *Like wind beneath wings the years have suddenly disappeared. How fast my children have grown.*

"You and your brother have inherited your father's blond hair and blue eyes. Neither of you inherited my brown eyes," I said.

She looked at me and smiled. "Jerry has Dad's blue eyes. I have his brains."

"The day your brother was born, all your Dad could do was tell the doctor and nurses, '*I sure made a pretty baby. I sure made a pretty baby.*' He was so proud. Then, twelve months later, you were born. Your dad brought me yellow roses." I took a deep breath and sighed as I remembered those years. "That was the first of many times your dad brought me flowers. Then he told them all over again, '*I sure make pretty babies.*' If I hadn't written names on your baby pictures, I wouldn't have been able to tell you two apart. You looked like twins. The nurse teased me, 'Carol, will you be back in another twelve months?' I told her, 'No, I don't have the energy for another one.'"

Janet changed lanes on I-95, and drove toward the off ramp and West Sunrise Boulevard, before continuing the conversation. "Of course, Mom, you had something to do with it. My friends tell me I look a lot like you."

"There isn't enough conceit left in southern Florida for anyone else. You and your brother have it all."

"I'm not conceited. I'm convinced." She laughed and took another drag from her cigarette.

"You know it was the light in your father's clear blue eyes that touched my heart. I fell right into them."

I became silent for a moment, thinking of the thirty-six years Alex and I had shared. *Our marriage has been good, except for the last few years. Alex drinks too much beer, and regularly falls asleep in front of the television. He is never loud or nasty and rarely complains. I wish he felt better. I feel such stress with bills and our tile business declining; it has had him in a depression for almost two years. He doesn't take care of himself. All he drinks is beer. Most of the time, he won't even eat.*

As for sex, forget it. I can't remember the last time we made love. I smiled, as I remembered he used to leave a screw on the *nightstand when he was in the mood. I remembered sometimes I wasn't in the mood. Alex would show me that contagious smile of his. He would put his arms around me and his fingers lightly stroked the back of my neck. I would relax and lay my head on his chest. Then we shared in each other's complete pleasure. How much I missed him. I wanted my husband back. To me, our sex was the fusion of God energy, Mother Earth, and Father Sun. It was the yin and yang that creates harmony and balance. We were supposed to grow old together. Finances can put such a drain on couples. We have fallen into a rut.*

Janet must have picked up on my thoughts. "I can sense the tension between you and Dad -- how about I buy you a vibrator for Mother's Day!"

Janet's remark didn't surprise me. We are very frank with one another. Isn't it amazing what your children tune into when you don't think they're paying attention?

"A lot of people use them," I laughed. "I'll keep taking my evening walks and cold showers. I just want to enjoy our day out, and not to worry about anything. The men need work jeans and I could use new bed sheets."

"I'm looking for pogs," Janet said.

"Pogs. You must have five albums of those fancy cardboard bottle caps. It's just a '90s fad. By next year it will be something else," I said.

"I like collecting them. It's real popular right now, here and in Europe. It's similar to an old game they used to play in Japan a few hundred years ago. Now they make them decorative and fancy. Lots of my friends are collecting them," she said.

The clock on the dashboard read 10 a.m. as we drove into the parking lot of the Swap Shop flea market. We were lucky to find a parking spot on the north side of Sunrise Boulevard; we didn't have as far to walk or cross the bridge on the other side of the highway. Several acres of stands, and hundreds of people were in front of us. A breeze carried the aroma of fried chicken and hot dogs drifting from the center lot where the food courts were located. We walked between crowds of people, ladies carrying shopping bags, children in tow holding tightly to their hands.

"It's going to be hot this afternoon, Janet -- I should have worn shorts."

"I know the food court has air conditioning. We can always stop and cool off when we need a break."

"Shop till we drop, I say. Just look at the variety of merchandise on stands before us."

CHAPTER 3

Danny

We walked toward the tent-covered stands. Janet complained of a rash and started to scratch her neck.

I looked at the red patches. "It must be a heat rash," I told her, although she'd never had a heat rash before. I had a weird feeling in the pit of my stomach, but I tried to ignore it. She walked in front of me. I watched her blond ponytail swinging. She was close to five foot five inches tall, and had a cute figure in her size eight brown shorts and white T-shirt. I had on blue baggy cotton slacks and an oversized white T-shirt.

Janet looked back. "Come on Mom, stop your daydreaming and catch up."

"We have the whole day to shop. I don't want to get tired before we even get started."

I hadn't used a scale for a long time. *I'll be fifty-six years old in June. For being only five feet four inches, I've really let myself go. I must weigh close to 160 pounds. I need some new energy, maybe a job outside of the house? I could buy an answering machine for phone messages.*

We walked halfway down the first aisle when I spotted T-shirts hanging across the front of a booth.

"Janet, look -- there's T-shirts with large mouth bass on them. Jerry and Henry would like those."

"We're here to shop for you, Mom. Forget the guys."

"Well, they do need work shorts, and I want to stop in the farmers' market across the next aisle before we leave today." We passed several novelty stands when I found a hairdini. The price was better than the home shopping channel on TV. The ad made it look so easy, the way they wrapped hair around a white bone hairdini into a twist. *What a break. It must be my lucky day.* A few stands down the aisle, Janet spotted pogs.

"We'll never get anything done now," I teased her.

"Quit your bitching. You got your hairdini." She laughed.

We had a good relationship. My children worked hard, and had a strong set of morals and common sense. I felt we were a lucky family. We had our share of problems, but we made it through them.

We approached the stand and started searching through the pogs. She picked up each one.

"Oh look, Mom. Here is one of Sleeping Beauty for my Disney collection, and one of the Cleveland Indians for Dad." Janet was excitedly looking through the bottle caps.

I looked for the vendor; he was behind the booth and had his back to me. He slowly and deliberately moved something on a shelf. Then he turned and made eye contact with me. I felt a surge of psychic energy deep in my root chakra, Kundalini. The charge of energy continued up my spine, through my chakras right out the top of my head. Shock! Struck by lightning? Was this a psychic attack?

Frozen in time for a moment, I couldn't speak. I could envision the look on my face. I knew this man! Danny Malone! I loved him. I could see his soul in the light of his eyes. It was spring, 1959. I was seventeen years old when I returned his engagement ring. I never thought the day would come when I would see him again. Remembering how angry he was when I broke our engagement,

DANNY

our eyes were locked, but I refused to speak. I just didn't know how he would respond.

I believed somewhere in my memory, my sister Emily had informed me that Danny died several years ago. Was I mistaken? How could he show up and materialize like this? He looked so alive. Did he fake his death?

The charge of psychic energy felt wonderful, like a free fall or falling from a bridge attached to a bungee cord. I hadn't seen or heard from Danny in thirty-six years. How could he just show up and tantalize me with this energy? I had a wonderful feeling. If there is a cloud nine I was on it. He definitely gave me a wake-up call.

My mind was spinning. Why was I drawn to this stand in this flea market today? What did he want from me after all of these years? Why now, when my husband hadn't been well? I wanted to call on my spiritual guides. How did he know I would be at the flea market today? How did he know Janet would stop at this booth to look at pogs? Had he been watching me? And if so, for how long and from how far?

A flood of feelings and memories from the past rushed through me. When we broke up I was seventeen and had just finished high school. I held a lot of emotional hurt and anger inside me. We never had a reconciliation or truce. Nor did we receive a healing from each other. Standing in the booth, he avoided eye contact with me after he gave me my psychic charge. What could he have been thinking? I saw a boy, about ten years old, standing behind the booth with him.

I did not want to make a scene in front of the boy or my daughter. I took a deep breath to regain my composure. Janet, apparently noticing nothing unusual, paid for a few pogs and we walked away. I was so occupied with my thoughts that I almost walked into several people and had to keep excusing myself. What had just happened to me?

Halfway down the aisle, I steeled myself not to look back. But

I couldn't resist anymore. I glanced over my shoulder to see if Danny was still there. He was standing there watching me as we left the aisle.

I was not thinking clearly, and things were a sort of a blur as we continued shopping through the flea market. I needed to get focused again and get back to my day with my daughter. Jewelry was always an eye-catcher. The scents of musk and sweet-smelling cologne saturated the air as we passed by the perfume stands. We searched through a table stacked with work shorts, looking for the right sizes. Then a stand with bed sheets caught my eye. The temperature started to rise from the afternoon sun. A mixture of aromas drifted from inside the food court area and piqued our hunger.

Stopping inside the food court for lunch and sodas, I asked, "Do you want pizza, chicken, or Chinese?"

"It's your day, Mom. I'm treating. You choose."

"Let's do Chinese." Janet ordered pepper steak for me and honey chicken with fried rice for herself. It felt good to sit and reflect a moment about what had happened. The cool air was refreshing inside the food court. As we ate lunch, I listened to the calliope music from the merry-go-round and watched the children standing in line, waiting for a train ride. I kept thinking about what I had seen and felt. Danny still looked good.

He must be fifty-nine years old now. His hair was black and wavy with some gray at the temples. His eyes were blue and he was not overweight. He stood about five foot nine and seemed to be in good shape. Years ago, when my friends couldn't think of his name, they would refer to Danny as "that good-looking guy up the street."

When we finished eating, I gathered the paper plates for the trash barrel and watched Janet prick her finger and go through her painful routine. She tested her blood sugar and then checked the readout. Pulling out her insulin pump, she pressed a button on the pump firmly twice. Janet didn't realize what had happened to me. I wasn't prepared to tell her.

"You are so lucky to have good health insurance where you work. That insulin pump has really helped you control your sugar," I said.

"I know, Mom. After all I went through in school, my pump has been a blessing. I can go anywhere and live a normal life. I just disconnect to go swimming." Janet smiled.

"I remember the day when you were in middle school; your blood sugar was low. The nurse from the school clinic telephoned me, and wanted me to come to the school because you weren't well. I told the nurse you might be having an insulin reaction, and to give you orange juice or a soda to drink.

"The nurse was really worried and wanted to call the paramedics. I told her I would get there as soon as I could. I didn't have our work truck. I got worried and then scared, wondering how I was going to get to the school. Fortunately, the principal called back and offered to pick me up. By the time I arrived at the clinic, you were sitting on the bed, your forehead covered with beads of perspiration. I sensed your fright and embarrassment."

"Mom, I know the whole school was watching me. I felt so humiliated."

"Yes, I know, Janet. You looked at me with tears in your eyes and said, 'Mom, I didn't even remember drinking the orange juice.' "

"The officer from the rescue squad told me that you had done everything right. I knew it was after one of those weekends you had played really hard. You always loved swimming, softball, and biking. You had a physically challenging weekend. You did the right thing when you told the teacher you felt dizzy," I said, and hugged her. "I wish you didn't have to deal with this."

Janet put her arm through mine and held on.

"Mom, I'm still upset when I have to depend on people. I want to believe I'm normal and healthy. You don't have to worry. I take care of myself."

"Janet, don't ever be too embarrassed to ask for help." I

was proud Janet was careful and successfully taking care of her disease.

We stepped outside in the bright sunlight to continue shopping. I could not get Danny out of my mind and was still feeling high from the psychic charge I had received. I felt intoxicated, high, and energetic. I remembered the excitement I felt when Danny and I when we were young.

"I'll catch up to you, Janet -- I want to check on something." I walked back past the booth to see if Danny was still there. But no, he was gone. What was I supposed to do? What was Danny into? *Maybe it was some kind of magic or cult activity, or was it a gift?* Just then, a white feather drifted down from above and landed on my shoulder. Tenderly I picked up the feather and placed it inside my purse. I searched for Janet. We bought corn and tomatoes, then headed for home.

CHAPTER 4

Returning Home

It was late afternoon when we returned home, hot, tired, and sweaty from the flea market. The air conditioning felt wonderful as we opened the front door. We had had a successful outing; even Janet's rash had disappeared.

The instant we got inside, we were greeted by our German Shepherd. Princess immediately started sniffing packages, looking for a treat. Luckily I had remembered to buy her a squeaky toy. Wouldn't you know? With that nose of hers, she found it.

Our husbands were hungry, so I made hamburgers and put a pot of water on for corn. Janet sliced tomatoes and placed plates on our white Formica-topped table. The boys took their seats around the table on chairs covered in a blue and white flower pattern. I turned on our ceiling fan to circulate air above the table just as they sat down. Princess stood near the table drooling. Then she lay on the floor beside Alex's chair. She looked at Alex with anticipation, her brown eyes meeting his blue ones, as though she hadn't been fed in weeks. Alex purposely dropped a hamburger on the floor for Princess. He spoiled her.

"This corn sure is sweet, Mom. You are such a good cook," Henry said.

"It must be from a local farmer -- the produce isn't always labeled at the flea market." I dropped a fork and reached to grab it before Princess found it. "Does anyone need salt?"

"Mom, sit down; you're worrying about needless things. We can get our own salt," Janet commanded.

"I have a lot on my mind and I'm distracted. Did you notice anything unusual at the flea market today?" I asked Janet.

"Just my rash -- that was out of the ordinary."

"The booth where we stopped to buy pogs, Janet -- do you remember anything?"

"I'm sorry, Mom. I just can't recall anything unusual."

After dinner, the men sorted and divided their new work shorts. Alex and the guys laughed and joked as they made an exaggerated fuss over their new clothes.

"Jerry, with your roofing job you'll have those new shorts looking like rags in no time," I said

Janet looked at Henry, "You're just as quick to wear out work clothes. Those construction nails always put holes in your pockets."

"Working in the sun the way you do, both of you will be looking older than your middle thirties in no time. Don't forget to use sunscreen," I said.

Janet and I cleared the table and washed the dishes. Hugs and goodbyes were next; then Alex took Princess outside to play in the yard.

After my children left, I sat on the sofa and leaned on the pillows. I couldn't keep from thinking about my experience with Danny's psychic attack. If he had the gift of a psychic connection while we were dating, why didn't he ever mention it to me? Was he afraid of my reaction? Did he think that I might not understand his gift? My mind was in a whirl again.

CHAPTER 5

The Phone Calls

The following morning, after Alex left for work, I sat at the kitchen table with my cup of coffee and toast. I had just opened the back door to let Princess outside. The telephone rang. I answered the call, thinking it was a customer of my husband's. There was dead silence on the other end. I felt apprehensive, uneasy...I held the receiver just long enough to figure out who it might be.

I said, "No more," hung up the phone, and went about my daily routine. I figured Danny was up to something. I did not trust him. I did not want any conflict in my marriage with Alex. After seeing him at the flea market he stirred hidden emotions in me I had long ago forgotten about.

That day, after my hang-up call, I kept having flashbacks from the past, feelings I had buried deep inside of me.

Danny had been in the Army, stationed in Korea, and came home on leave at least twice during his last year of service. After my high school classes, Danny would pick me up in his blue Plymouth convertible with a checkered ragtop. During my last study period I was always anxious waiting for the bell to ring, when I knew he would be waiting for me. After school when I stepped out of the front door with classmates, the girls' mouths dropped open when they saw Danny standing beside his car, wearing his brown Army uniform, holding the passenger door open. Danny walked with controlled heavy footsteps and always seemed to be in a hurry. He chain-smoked Marlboro cigarettes and was so sexy the way he placed a cigarette between his lips. I waited for him with anticipation, my heart pounding.

In study hall, I would write on my steno pad: Mr. & Mrs. Danny Malone, or Carol Malone, or Mrs. Carol Malone. I wrote over and over again. For three years, I had been dating Danny. Every morning, I would meet his sister Irene before school; then later we would have lunch together. Danny's mother Margaret seemed to be kind and easygoing. She had to be, raising his three sisters and three brothers. Danny's stepfather was employed at the steel mill, but work was erratic. The family was considered lower middle class, so they didn't have a lot of material possessions.

I remembered one day that I was invited to dinner at Danny's home. Spaghetti was on the menu. It was served with tomato soup as a sauce. I imagine Margaret would have preferred to make real spaghetti sauce, but I didn't mind. The family sat together at the dinner table, passing their plates for food. The yellow Formica table was faded and worn from use. The feeling I had was warm and friendly.

While remembering Danny, I now felt a touch from Spirit. A warm energy expanded from the center of my stomach into my heart.

My mind flooded with a memory. It was an autumn evening, Danny and I strolled in the neighborhood park. The park seemed deserted, except for a few people in the distance. I felt a cool crispness in the air, forecasting the coming of turkey, crystal snowflakes, and angel messengers -- the beginning of winter.

Leaves were turning red and yellow. I knelt and picked up a few colorful ones off the ground. Danny gathered a handful of leaves and tossed them at me. Laughing, I reached for and ruffled his black curly hair, the curls that refused to obey his command.

I stood up and Danny stepped behind me, put his arms around my waist and kissed me tenderly on the neck. I turned and looked into his blue eyes. I felt a gravitational power within me. We kissed passionately and feverishly. His hand found the opening in my jacket and slid toward my breast. I fought with myself over the shiver of desire I felt. Then, I inhaled a deep breath of fresh air, gathered my strength, and pulled away.

"Take me home," I said.

I was taught to be true to myself. I wanted to wait for marriage. Everyone in school knew the girls who would let the guys go all the way.

I turned on the path toward his car when he placed his hand on my shoulder and drew me close. I felt his fingers gently caress the back of my neck. My spine tingled. I felt the warmth of his body, and his breath. He whispered in my ear, "Nobody else will have you. You're mine."

A few months later in my home, I met him at the front door. He wore a suit and tie. "You look terrific in your black dress," he said. "The red sleeves show it off nicely."

Danny embraced me and kissed me fully on the lips. Together we stood with our arms around each other and admired the decorated pine tree in the living room. It was Christmas Eve. The tree lights seemed brighter than usual. He put his hand in his pocket, and pulled out a small blue velvet box. He opened the box and gingerly removed the contents. I saw a beautiful diamond ring. He placed the ring on my finger.

"Will you marry me?"

My heart skipped a beat. "Yes."

I threw my arms around him. "I will always be yours."

When I returned to school that January, I showed off my engagement ring to classmates. Many of the girls in my senior class had received engagement rings that year. I thought Danny, and the ring he gave me, would provide me with commitment and security for the future. I used to wait for the mailman, wanting another letter from Danny in Korea. I was excited and looked forward to my graduation in May, only six months away.

The following spring, just before graduation, I suddenly had the feeling that my whole life was before me. Some of my classmates were planning their weddings, and others were anticipating college classes and professions. I began to realize that I wasn't ready to be tied down to marriage, dishes, and diapers. I wanted to discover the world, travel, meet new people, and do wonderful things. I didn't know how to tell Danny how I felt, but I knew we were always truthful with each other.

"When we are married it will be you and me. We will have a new life together," Danny told me, giving me visions of a small apartment with pretty ruffled curtains, inviting friends over for dinner. But then I thought, *No, I need to see other places*.

One night I was in our neighborhood taking an evening walk when I had a vision of Danny. There was a dark shadow with him. Somewhere from deep inside me came a thought, *Yes, I used him*.

Did I use Danny? I had to ask myself.

All of my life, Mom and Dad had planned my time and events. School gave me a time to be there and a time to leave. Tests were scheduled. Church on Sunday and family reunions...I was always told where I was supposed to be, and what to do. I was graduating and I had no direction or plans. My parents had no money to send me to college. I thought Danny would take care of me. We could work and build a life together. Then, I had a feeling I would be stuck forever in a place I couldn't get out of.

Now I knew I didn't want to get married. A twinge of fear filled my solar plexus.

After being discharged from the Army, Danny needed a job. A few days later, my Aunt Ruth told me she would give Danny a reference to help him seek employment in the rubber factory where she worked. On the second day, when Danny was supposed to work, Aunt Ruth telephoned my mother.

"Marge, I gave Danny a reference so he could get a job at the factory. He didn't call or show up for work today. This is going to make me look bad."

"What?" she said. "I'll certainly say something to Carol."

When my mother gave me the news that afternoon, I walked down the street to Danny's mother's house. I was upset. If he didn't like the job, he could have told me. I didn't want Aunt Ruth to be embarrassed. I found Danny in his room, lying on his bed. He smelled of stale beer. I stood at the door clenching my hands into fists.

"Danny!" I said. "Get up!" I said, seething. He rolled over and looked at me. I could see he was drunk. "Did you work today?" I asked him.

"Yes," he answered. Then he looked away from me.

I knew it was over. I removed the engagement ring from my finger and handed it to him. He refused to take it. I couldn't believe he had lied to me. What was he thinking? I knew it was no use arguing with someone when they were drunk.

I left his room and found Danny's mother standing in the kitchen, I handed her my diamond ring with tears rolling down my cheeks.

"Carol, don't leave, stay so we can talk." She was distraught.

"I can't stay, I need to leave." I wanted to stay but I knew I had to go. With tears rolling down my cheeks and my throat constricted with sobs, I left. What was I going to do now? I was almost home when I heard a crash. I immediately had this fearful feeling.

A few minutes later, my next-door neighbor, Jane, knocked on my front door. "Danny jumped into his car and drove into a telephone pole. He totaled the Plymouth."

I wanted to call and see if Danny had been hurt, but I was frightened. I knew he was drunk and angry. He crashed his car. What would he do to me?

Later that evening, Jane phoned me.

"I called Irene and asked her what happened. She told me he wasn't badly hurt."

"Thank you, Jane." I let out a sigh of relief. "Thank God he is okay. I'm so upset right now -- I'll call him tomorrow morning when I've calmed down," I said between sobs. "I'm so confused and angry. Thank you for helping. Goodbye."

I knew the whole neighborhood was watching and talking about us.

Danny called me the following morning.

"If I can't have you, Carol, no one can. I'll kill you."

I hung up the phone. The tightness in my stomach and lump in my throat brought tears to my eyes. I didn't believe Danny would hurt me. He was just angry, spouting off words he didn't mean. I loved him. Why would he hurt me? I was confused and scared. Dear God, I had never seen Danny this angry! What if I had married him, and then found out what his temper was like? When we had small children, I would have been walking on eggshells, not wanting to piss him off.

I loved Danny, but he needed time to cool off, and I needed to make some constructive decisions about my life. I had to get away.

In the classifieds section of the newspaper I found an ad for airline school. That would be a wonderful way to work and travel I thought. I applied to Grace Downs School in New York City for training as an airline stewardess. At the beginning of August, I packed my suitcase, preparing to leave for New York City. My Aunt Fay gave me a navy blue suit-dress for my trip. My middle sister Emily loaned me a thousand dollars for airline school tuition. The following Monday morning, my parents drove me to the train station. They embraced me.

"We'll miss you, Carol. I want you to be careful in the city, and don't go out at night by yourself." Mother smiled, and then she handed me an autograph book to read on the train.

I felt overwhelmed, and ruffled the pages. "Thank you, Mom. Once I made my mind up to leave everything happened so suddenly -- I'm surprised you had time for everyone to sign my book."

"Friends stopped by the house, and your dad drove me to the others."

"Yes, your mom keeps me busy." I hugged my dad and mom and stepped toward the train with my suitcase, not sure what I was in for, but happy to be going. After I settled in my seat and nodded a greeting to passengers close by, I relaxed and opened my book.

On the first page I read,

"I'll write on the cover to save room for your lover.
Your five-year-old nephew,
Mark."

Turning a few pages I found,

" 'Tis the middle of the book I chose as I haven't the most or the least to say. I'm not the oldest or the youngest in our family, nor the richest or the poorest, but always close to your heart I'll be, whether it's happiness, trouble, worries or triumph, health or success. I'll always be by to smile, encourage, and cheer you on.
Your sister,
Emily."

I found a page my Aunt Fay signed.

"All of the luck and happiness in the world."

I was surprised to find a note from Danny's sister.

"To Carol, one of the nicest girls I've known for a long time. I hope someday we will be close friends again, as we were before. Best of luck in your career, and I hope you have all of the success possible.
Love always,
Irene."

LOVE FROM THE OTHER SIDE

There was an Air Force officer sitting across from me. I thought he might have noticed a tear in my eye as I closed my autograph book, or maybe it was the somber look on my face, because he invited me to lunch in the dining car.

His invitation was comforting and I gladly accepted. I ordered a chicken sandwich and a Pepsi.

"I'm going home on leave," he said.

I said, "I am going to school in New York City." I looked at him and said, "I have never traveled alone."

He smiled and said, "You'll be fine."

I thanked him for my lunch.

After our lunch, I opened my book and read a few more messages from friends and family.

My Aunt Sue:

"Carol, as sure as your wedding day does come, a broom to you I'll send. In fair weather use the bushy end. In storms, use the other end."

Another note read:

"My brother took the front so I'll take the back & be the last to wish you cheers, happiness, health & guidance for your future.
Your niece,
Sandy (3 years old)"

I skipped to the last page and found,

"Be it ever so humble, there is no place like home.
We'll miss you,
Mom and Dad"

THE PHONE CALLS

I didn't get to finish reading my autographs. I began to doze off.... Before I knew it, four hours had passed and my train had arrived at Grand Central Station. I was amazed, looking at how big it was, and tried to make my way through the crowd so I could hail a taxi. The taxi driver was middle-aged, wearing a short-sleeved shirt and baggy pants. He was all business as he asked for addresses and destinations, and had four other fares in the cab. We stopped at two big hotels to drop customers off, before I reached the front doors of Grace Downs School.

New York was nothing like the quiet country roads of Ohio. The tall buildings and narrow streets were overpowering. I couldn't get over the heavy flow of traffic, and honking horns from cab drivers, or the sound of screeching brakes and buses. I knew the underground subway would be a new experience for me.

Once in the school, I found the reception room was large. I stepped on the gray tile floor (or was it marble?) and noticed a winding staircase. There was one elevator. I was informed it was out of order most of the time. I had to walk up three flights of stairs to my dorm room. The girls in my dormitory were friendly and we had fun getting acquainted. They were all just graduated from high school, slim and pretty. We told stories of family and romances, and then, lost loves. One of the girls, Sue, told us, "My boyfriend told me if I leave home, it's over. Well, here I am."

In the evening I gazed out of the dormitory window, and stood in amazement when I saw the color of a brilliant red sky against the tall majestic buildings. We found a restaurant close by the school where we could all eat together for a reasonable price. Saint Teresa's Church was just across the street from school.

While I was at school, Danny wrote several letters asking me to come back. I knew if I went back he would not have changed. One day his temper would explode, and I didn't want to deal with it. I did not encourage Danny; I wrote back and told him to go on with his life. I stayed in New York for two years.

After graduating from airline school in November, I completed several applications only to finally find out that I had to be twenty-one to be hired, and I was still only nineteen. I was full of excitement when found employment with a car rental agency. Their office was in Idyllwild Airport. There was a nice Italian family who had a room for rent in Queens. I felt lucky that I found them and I felt safe there. Emily was kind enough to loan me more money. I stayed in New York until I paid her back. Then, during the summer of 1961, I returned to Ohio. I had become lonely living and working in New York. I missed family. I enjoyed my newfound friends, but I never felt connected. I felt alone and detached. They were not like family. I was homesick.

On my second day home Jane told me, "Danny is married. He was married right after Christmas last year, in January of 1960."

Wow, I thought. *It was the fall of 1959 when Danny was asking me to come back to him.* On Christmas Eve, I flew home from New York to be with my family. I entered the house and had hardly put my suitcase down when I heard the phone ring. Danny's mother was on the phone. She was hysterical, sobbing and begging me to come to her house. I wondered what things he had gotten into. Something wasn't right, but the feeling wasn't clear, and neither was she.

My high school years hadn't done much to prepare me for some of the sensations that were part of my life. I knew nothing about anything resembling magic, black or white. Six years in parochial school were strict and conforming. Ask a question and get your hand slapped. In my innocence I looked for good in all people. I had always felt a draw to Danny and Irene, like a magnetic force. Was it Karma or something more? I didn't know. I had never seen anything that had any indication of the occult in Danny's or Irene's home.

I had discovered Danny's temper when he wrecked his car. This call from his mother sent a chill that was completely wrong. Now I felt an inner knowing that Danny had secrets he didn't share with me.

"I'm sorry, Margaret, I just can't come back. I will not let myself get involved again. I have to go back to New York." I hung up the phone.

I felt deflated. My ego was hurt. I felt a churning in my stomach. After being with Danny and his family for several years, I had feelings -- maybe I loved him. I had hoped Danny would straighten out his life, take control of his temper, and find a positive outlook toward his future. The one man I cared for stopped waiting for me and married someone else. There was an omission of truth in our communication with each other. My reality was that I knew he would never change; he was a troubled soul. Why, I wondered, did he marry so suddenly? Was she pregnant?

A few weeks after returning home, I found a job. Working the afternoon shift in a box factory killed my social life, but I needed the job.

Now, thirty years later, I was sitting on my couch in Florida and thinking about Danny. This heavy weight, this thick gray energy, was still between us. I had a feeling the message while I was in the light, in my dream, only a few weeks ago, had a lot to do with Danny. All of these events seemed to come together: seeing Danny at the flea market, my hang-up phone calls, and my feeling of being disturbed.

CHAPTER 6

Alex

I poured myself a glass of iced tea and sat at the kitchen table. Princess lay on the floor at my feet, her squeaky toy safely between her paws. I glanced at the kitchen clock, and couldn't believe that half of my day had evaporated into thin air. With my daily calendar in front of me, I began to check my appointments.

Then without effort, my conscious mind drifted back to the years when Alex and I had our first encounter. It was in the fall of 1961. A neighbor had introduced me to my future husband, Alexander Shimp. A tall six-footer, he had blond hair and the most beautiful blue eyes. When I looked into his eyes, I knew, somewhere, sometime, I had been with him before. *Was it in a dream, or was it a past life?*

When I met Alex, he had six months left in the Army. His uniform showed off his broad shoulders and slim waist. On Monday after our weekend date, he left for Cleveland, where he was stationed. He came home on weekends, and we always spent time together as we learned about each other. Alex would pick me up in his 1955 green and black Plymouth Belvedere. We would go to the movies or visit friends.

"Would you like to go to the movies next weekend?"

"Sure, what's playing?"

"I'd like to see *The Guns of Navarone*."

"It figures, you want to see a war movie. I would like to see a love story. Wasn't *West Side Story* advertised in the paper? The reviews I read were excellent."

"A musical -- I don't know, all of that singing and dancing."

"It has high ratings and the script tells a good story."

"Okay, *West Side Story* it is. Because I like you," Alex winked.

Alex told me he sang in the officers' club. When he sang to me, he sounded a lot like Johnny Mathis.

"*Until the Twelfth of Never. You ask how much I need you. Must I explain. I need you, oh, my darling, like roses need rain.*"

Alex could dance. He loved to dance. In my parents' kitchen, while I made a pot of coffee, he held out his arms and pulled me close. I inhaled the aroma of his Old Spice aftershave and we stepped to the rhythm of his song. Even when he walked, it was with style and grace. We discussed our experiences in Catholic school, going to Mass every morning and receiving communion, and he would reminisce.

"One day in fifth grade I was so bored in class I was doing something distracting. Not sure what I was doing, but I remember standing in a corner of the classroom for an hour. It seemed like eternity. Ha, ha, boy do I remember when the sister caught me with my hand in my pocket. She whacked my hand with a ruler until I thought I would never be able to use it again."

My contributions to the conversations with Alex were usually simple and pleasant memories. "I remember on cold winter mornings when we couldn't eat breakfast before communion, mothers would serve hot chocolate in the cafeteria after mass. The warm liquid tasted and felt so good as it filled my stomach."

The sharing helped create a bond, and the relationship grew. One evening returning home after a night out with my girlfriends, I found Alex setting on my front doorstep.

"What are you doing here?" I was surprised.

"Waiting for you," he said with a positive tone.

"Were you worried about me?"

"Of course -- I don't want you to hook up with anyone else."

I just might have to marry this guy.

I was still working the afternoon shift in the box factory. I did not like the job. The heat coming from the presses was stifling. Stripping boxes, I had paper cuts up and down my arms. I quit right after New Year's to look for other employment.

On Valentine's Day, Alex gave me a gold heart-shaped box of chocolates. When I opened the box I found a beautiful solitaire diamond ring. I threw my arms around him and kissed him.

"You know, Alex, it wasn't long ago that I asked my Mother how she would like blue-eyed, blond grandchildren."

Somewhere in my mind, subconsciously, I knew our souls were meant to unite.

He took the ring and placed it on my finger.

"It fits perfectly! How did you know my size?"

"Your mother told me."

My parents, his parents, and Alex and I went out to dinner. Alex had told me that his birth mother died when he was twelve. His stepmother was Baptist. His father didn't seem to care about religion. Everyone ordered fried chicken and coleslaw. It was a family-style restaurant and the waitress placed a large bowl of mashed potatoes on the center of our table.

Alex and I attended the premarital classes in the Catholic church, which were required to prepare for marriage. The church also posted wedding banns in the bulletin three weeks before the wedding date. The announcements were meant to notify everyone about upcoming marriages, so that anyone who knew any reason that should prevent the marriage had time to come forward.

Alex's stepmother Ann and my mother gave us a wedding shower. If I remember right, we unwrapped three coffee pots.

On the thirtieth of June, 1962, we were married on a Saturday morning in Saint Paul's Church, at a High Nuptial Mass. We had a large Polish wedding. My mother and I mailed over 300 invitations. I invited my ex-fiancé Danny's mother Margaret. I wasn't sure if she would come. Danny was already married, but I had known Margaret since I was a freshman in high school. I had hoped there wasn't any tension between us. I hadn't kept in touch with Danny's sister Irene when she married and moved to another state.

The wedding was costing Alex, my parents, and me around two thousand dollars. So when my cousin offered to loan me her wedding dress, I accepted.

"It's a beautiful dress, Carol," Judy said. "I paid seven hundred dollars for it."

I was grateful for the offer. It was truly a Cinderella dress. It had a sweetheart neckline and long sleeves, and a full floor-length skirt with a chapel train. The skirt fell in tiers of lace, embellished with pearls and sequins. I bought a veil and tiara covered in white satin with crystal teardrops. My cousin Rose, and Alex's sister Marie wore knee-length blue lace dresses, as my maid of honor and bridesmaid. The men wore tuxedos. Mother was elegant in her beige eyelet lace over satin dress.

I was so pleased. "Dad, after seeing you for all of those years in a hard hat and work clothes, carrying your lunch bucket to the steel factory, you look real handsome dressed up."

"Well, you're my last daughter to be married. This is the last time you'll see me dressed up." He smiled.

I did see my father one last time in his suit, as he lay in his coffin when he passed away in March, 1987.

In church, before Mass, we had placed white bows on the edge of the pews. The florist brought white mums, blue daisies, and white roses for the altar.

As the organ music began to play "Oh Promise Me," Rose handed me my grandmother's lace handkerchief. I held my bouquet of white roses, mums, and blue daises. My father placed his arm in mine and we started down the aisle. I noticed Margaret sitting at the end of a pew by the center aisle. My eyes welled with tears and I cried quietly all the way to the altar. I had an uneasy feeling in the pit of my stomach. I felt like Margaret was not there to celebrate my marriage to Alex, but as a warning. I was in sixth grade when I met Irene, her mother Margaret, and their family, so I felt close to them; they were my second family. I saw them almost every day after school and on weekends. When I broke up with Danny I had this emptiness inside of me. I wondered, *Is there an off switch to lessen hurt for the heart?*

My father gave my hand to Alex. Then he stepped back.

With my eyes half-closed, I felt the back of Alex's hand tenderly wipe a tear from my cheek.

"You will be happy. I love you," he said.

I noticed a stream of bright sunlight pass through the stained glass window of Jesus above the altar. Immediately, I felt blessed. My tears dried and I smiled. Alex healed my damaged heart when he wiped away my tears with his hand. My spirit was embraced with fulfillment. We repeated traditional vows.

We rented the dance hall in the Polish club for our reception. It was six p.m. when I stood at the entrance of the reception hall with Alex behind me. I noticed a broom lying on the floor across the doorway.

I picked up the broom. "Who left this here for someone to trip over?" I complained. I heard laughter.

Then my mother told me, "It's a Polish tradition. Because you picked the broom up, it is said you will be a hard-working and caring wife."

"There is a lot to be said for tradition. I won't let you work too hard," Alex said grinning.

My aunts had been busy planning the menu to fill tables with food. My Aunt Sue made decorated cakes for a living, so she had offered to make my wedding cake. My sister Emily made nut rolls and cutout cookies in the shape of bells. My mother made her famous pierogies filled with potatoes or sauerkraut, and then boiled and fried. From somewhere there was a roasting pan filled with stuffed cabbage rolls, and chursciki (or Angel Wings), a light waffle-type cookie sprinkled with powdered sugar. By the time the tables were full of food, the reception hall began to fill with guests. I did not see Margaret among them.

Alex had hired a polka band. The traditional dance is a fast two-step. It is music that warms the heart and is contagiously happy. Once the music starts, your feet won't be still. Alex held me in his arms and we danced the first waltz. The band played many of the traditional songs, "Hoop-Dee-Doo," "Just Because," "Tic Toc," and "Blue Skirt Waltz," to name a few. My brother-in-law Larry (married to my elder sister Vivian) and Alex's brother Jesse tended bar. We sat at the bridal table watching our guests dance. Glasses of champagne were passed. Alex and I stood and raised our glasses to a toast.

"Zostan Z Bogiem!" The guests toasted, which translates, "The Lord be with you."

Everyone began to dance and the band yelled, "E -Ye -E -Ye- Oh."

And the dancers yelled back, "E- ye -E -ye - Oh."

Watching my mother and father on the dance floor gave me a warm feeling -- Dad in his tux and Mother in her beige eyelet and lace dress. They were a vision of their younger years, twirling in step to the music.

We sat at the bridal table and I noticed a mist by the hall door. I had never met my grandmother on my mother's side. She died when I was very young, but I had seen pictures of her. I noticed a lady standing by the door that looked just like my grandmother's picture. There was a mist around her head and shoulders. Her hair was combed back in a bun and she wore a long black dress. I smiled toward her and she smiled back. I sent a thought. *"Welcome."* Then I saw her make a gesture with her hand; it was a blessing.

ALEX

Someone called for the bridal dance. I stepped to the center of the floor, holding a white satin purse with strings encircling my wrist. Guests lined up with money in their pockets. Alex took the first waltz, and then my father tapped his shoulder. Then, one at a time, the other guests took a turn and cut in, each guests placing money in my satin purse. Everyone clapped.

A short time later, I sat on a folding chair in the middle of the dance floor, and waited for Alex. Like a warrior being knighted, he knelt in front of me. His smile was contagious. Our eyes met, and my heart skipped a beat. I had saved myself for this moment. It was the light in his blue eyes. I could almost see the children we would have. He carefully lifted my gown above my knee while everyone ooh-ed and ah-ed. He removed the blue and white garter from my thigh, stood up, and tossed it to the single men. His best man Mike caught it. Then I threw my bouquet, and my ten-year-old cousin Gail caught it.

Our photographer asked us to have a picture taken with handcuffs on. We did. When we wanted to take them off, someone had lost the key.

"You'll never get away now," Alex teased.

"You planned this." I started to giggle because I knew they had pulled a fast one on me. I looked at Alex, and then at the photographer.

Then guests began moving chairs, looking under tables, and checking their pockets. After a few minutes with everyone pretending to search for the keys, the best man came clean and handed the keys to us. There was laughter.

Before we left our wedding reception, my mother stood me in front of everyone and removed my veil. She then placed a blue and white hostess apron with ruffles around my waist.

"This symbolizes that you give up your innocence and accept your duties as a wife, a hostess, and a mother." The elderly women, some from the old country, clapped.

"But, Mom, I'm a liberated woman -- we don't have to be slaves in the kitchen, in this day and age," I said.

"Liberated or not, there is still laundry to be done and dishes to be washed."

"I'm a good dishwasher," Alex smiled.

We said our goodbyes, and as we walked toward to door, I heard the band play: "*We don't wanna go home, the party's just begun. Save the band just one more drink. We don't wanna go home, we're having too much fun.*"

Alex had recently been hired by a canteen company refilling vending machines, so two days was all the time we had to spend in a cottage by Lake Erie. We left for our honeymoon in Alex's green and black Plymouth Belvedere. We drove a few miles on Route 62, when we discovered smoke drifting from under the hood of our car. Alex stopped on the side of the road, and when he lifted the hood, he found fireworks attached to the motor. He quickly grabbed a rag from under the front car seat and removed the fireworks. We sat there on the side of the road until the car engine cooled off.

I was seriously upset. "Oh my God, Alex, we could have been killed. The car might have blown up."

"I think I know who did this," he told me. "I'm going to have a talk with them the first chance I get."

I wonder if Danny put fireworks on the engine? It was someone's bad idea of a practical joke, we agreed. We hoped there were no other surprises in store.

We were tired after a full day of events. I slid across the front seat close to Alex. He placed his arm around me and drove with one hand on the steering wheel. It was around three a.m. when we arrived at our small two-bedroom wooden cottage. We slept late the following day and enjoyed our privacy. Then we bought a chicken dinner in town, and held hands while we took an evening stroll by the lake.

We returned to my home the following Sunday. My mother invited the family over to watch Alex and me open our wedding gifts. She had crocheted a beautiful blue and yellow afghan for our gift. Emily gave us a beige satin bedspread. My elder sister

Vivian and her husband Larry were there and gave us two beautiful Italian wall plaques with cherubs on them. Alex's mother had gifted us with a dinette set.

I was happy Vivian and Larry could make it to our wedding. They lived in Chicago and didn't come home often, because Larry couldn't get off work. The conversation over the unwrapping included parts of the reception we had missed. It was then we heard stories of my two uncles and a bit too much alcohol. After we left our reception, the two of them created such a problem that the police had to be called.

Later that evening, Alex and I left that happy family gathering to spend our first night in the little apartment we had found on the second floor of an old home. We already had our bedroom set and the kitchen was fully furnished. We would worry about the rest later.

In due time, we decided we would purchase my parents' home when they planned to retire in southern Florida. Alex left the vending machine company to work for Republic Steel. After he was laid off from the steel factory he applied for work at the post office. Our first child, Jerry, was born on the third of June, 1963. Janet made her entrance into the world on the twenty-fifth of June, 1964. I remember going to confession a few days before I went into labor with Janet. I kneeled in the confessional on the wooden platform and made the sign of the cross.

"Bless me, Father, for I have sinned."

At the end of my confession my priest asked me, "Are you going to take birth control?"

"Yes, Father."

"Then I cannot give you absolution."

I left church with a hole in my heart. God knows we could not afford another child. Having two babies in twelve months was finically strapping.

LOVE FROM THE OTHER SIDE

When I told my mother what happened she said, "The church will tell you to have them, but they won't help you raise them."

I understood my mother's bitterness. Years before, my father's sister had passed away after having her seventh child. My mother had knocked on neighbors' doors asking for clothes for the children, trying to help, but the times were just too difficult. In a depression with no work and no food, finally my uncle was forced to come to my mother, defeated and desperate. My mother accepted the oldest child into our home, but the others were put in an orphanage. It had to be a heartbreaking situation.

Absolution aside, we baptized our children in the church where Alex and I were married. I wanted my children to have a spiritual background and to learn about God while growing up. Hopefully, they would have more support, information, and understanding than I had growing up. I always felt a connection to the church but still, some things I kept to myself. On three different occasions, while sleeping, I had visions of the spirit of receiving communion. Upon awaking I clearly remembered the sensation of opening my mouth and a wafer placed on my tongue.

By 1967, we had lived in our house for five years. Things were still tight; there wasn't extra money for social events. Our savings amounted to nickels and dimes placed in a pretzel jar. One day in March, I was looking out the kitchen window while washing dishes after lunch. Snow in our back yard was starting to turn into slush, dark holes turning black and gray. I suddenly felt a chill in the house. In the basement I checked the gauge on our oil tank -- it read empty. It was forty degrees outside and we had no heat. I immediately telephoned the oil company.

"I paid my bill late, but you could have warned me if you weren't going to bring oil. I have two small children, a three- and a four-year-old, in the house," I complained.

"Sorry, we have no oil to give you. I can't promise any delivery until Monday."

"It's only Thursday and forty degrees outside. Thank you God, it's not freezing.

What are people supposed to do?" I asked.

"We will send a truck out as soon as one comes in. That's all I can tell you." The woman hung up the phone.

A feeling of conviction, like the undercurrent of a river, pulled at me. I had just had enough of this, and knew something had to change. I didn't want to live this way anymore. I needed sunshine to lift my spirits. The children and I were stuck in the house for days on end. If I needed to keep our car to shop for groceries or take the children to the doctor I had to worry about dropping Alex off at his work site and picking him up later. The stress and restrictions were oppressive.

That evening we sat at the table with sweaters on while we finished dinner. I thought about the trip we made to Florida two years before. *The ocean was beautiful; I was amazed at the blue-green water that held so much quiet power. I fell in love with the palm trees.*

"This is crazy," I told Alex. "You know, my parents moved to southern Florida five years ago. I think we should sell the house and move."

"I don't want to move," Alex said.

"Alex, I need a change. If you love me, please think this over." I looked at Alex with tears in my eyes. I couldn't believe his directness. "I don't ask you for much, but I really want this. I'm hoping for a better life and a future with more choices. Alex, I know a change might not work out for us, but I need to try. The children would be able to have a relationship with their grandparents. You wouldn't have to worry about going to work in the ice and snow. You always complain about how dangerous it is when you're climbing up those steep icy steps on Folton Road. And about how often your mail truck gets stuck in snow and you have to sit in the cold waiting for help."

"Do what you want, Carol. You will, anyway. I'll go along with whatever you decide."

"Alex, if they ran out of oil this year, what will happen next year? The whole north could freeze. Think of what the sunshine will do for the children. It would be better than being closed up in the house all winter."

I had made up my mind. That year, I started having garage sales. Within a year, we put the house up for sale. I kept telling Alex to think of the money we would save on snow tires and winter coats, shovels, and heat in the winter. The house finally sold in May.

Alex's stepmother was in tears. "I'm going to miss the children."

"You can come visit us. Think of a Florida vacation."

All through that year, even with the garage sales and signing the house with a realtor, Alex remained hopeful that I would change my mind.

"Ohio looks good in the spring. I enjoy the change of seasons." He looked at me. "You know my parents don't have the money to travel to Florida on a whim. We won't get to see them very often." He looked at me with his liquid blue eyes. "I guess I have to make the best of a bad deal."

"Alex, I'm sorry I can't fix all of these problems, but I want a change. I feel stagnant here. We can make trips back home and they can visit us. Can we have a truce?"

"You know I love you. If that is what you, want we will make it work."

In May of 1969, we packed our furniture in a U-Haul trailer behind our seven-year-old blue Pontiac Tempest. We put our children, Jerry and Janet, in the back seat of our car. There were no seat belts at that time. They played games on the way to Florida. They counted red Chevys, to see which one could reach one hundred first, or they looked for signs to find letters of the alphabet. In between, they colored in books. Once in a while, they gave Alex and me some peace by falling asleep. We told stories of sand, beaches, and dolphins.

After two days and one night in a motel, Alex parked our car in my parents' driveway. Mom and Dad were waiting on their front porch; hugs and kisses were abundant. When I saw the palm trees and we drove by the ocean I felt a warm glow in my heart. *Dear God*, I thought, *how blessed I am*. Our son Jerry chased geckos. He was determined to catch one to keep for a pet. Janet was romanced by the flowers. Grandma had these beautiful white sweet-smelling gardenias in her backyard. I couldn't get over the sunsets. Large white cumulus clouds were surrounded with colors of blue, gold, and orange. We would make it through hurricanes and storms. If we lost power, we could play cards by candlelight.

We hadn't been in Florida one week when Alex was reinstated by the Postmaster. My parents were so happy to see us. They had a chance to spoil the grandchildren. Southern living worked at first. However, after living in Florida for a while, Alex began to complain almost every day about the heat. When he came home at night, his uniform shirt was wet and white from salt. He was exhausted. "Where will we go when I retire?" he would ask. "We're already here."

In September, Jerry started first grade and Janet started kindergarten. They had no problem making friends. Alex and I spent many evenings playing pinochle with my parents. Alex and my father were always bidding against each other. Alex would usually end up giving in. The children played with their friends until it was bedtime.

A year later, we found an apartment close by. During the next few years, we lived a normal life in Florida. Janet and Jerry made their first communion and confirmation in Saint Henry's Catholic Church. Alex left the post office and used our small savings to purchase a tile franchise. A new patent had just been released. It was a coating covering cement on pool decks and driveways. On a wing and a prayer, he had dreams of building his own business.

Alex liked working for himself. The hours were long. He labored steadily and worked hard. The competition was against him. He didn't have enough collateral for a bank loan. There was no one to financially back him.

LOVE FROM THE OTHER SIDE

By the time the children were in high school, we stopped going to church. Alex and I were both working, time was precious, and there was always so much to do.

By 1972 I worked evenings in at Madar's bingo hall on Federal Highway. Many evenings, Mother would go to work with me so she could play bingo. The little money I made bought groceries. I felt better because Alex could be home in the evenings with our children. You never knew what kind of trouble teenagers might come across.

I would come home from work around eleven p.m. In the mornings, I would wake up with Alex. He left early when he needed to pick up tile and material for a job site. Then I prepared the children for school. Boy Scout and Girl Scout meetings filled any extra time. Sundays were the only mornings we could sleep in.

༄༅༄༅

I glanced out of the window at the palm trees in our yard. Seeing Danny at the flea market had triggered my memories from so long ago. My parents had passed on and Alex's stepmother and father had died. Jerry was now working and Janet was married. I wondered what to do about these hang-up phone calls. I decided to tell Alex about my strange encounter at the flea market.

༄༅༄༅

That evening, Alex came home from work early. We had dinner. When he was relaxed with a beer in his hand, I told him, "I ran into Danny, an old flame of mine from high school at the flea market, and I had a hang-up call this morning."

"Don't be ridiculous, after all these years?" He laughed.

I did not laugh.

Alex thought for a moment. "Hey, that's the guy who picked a fight with me, back in Ohio when we were dating. I was in one of the neighborhood bars when he came up to me and said, 'That's my girl!' He threw a punch at me!"

"I remember now, the time you told me. Why did he care? He was already married." I thought a moment, and then smiled. "Who won?"

"I had back-up and he left. What the hell is he doing in southern Florida?"

"I don't know. Maybe he has a weird way of saying hello."

I said nothing more to him. The hang-up calls continued at the same time every morning for a week. I wanted to talk to Danny, find out what he was thinking, and try to bring closure to my past. I thought Danny was playing head games with me.

The phone calls stopped for a week. Then they started up again, different times and different days. I knew it was Danny calling me. There was no one else. Danny always seemed to know when I was home alone. One afternoon I answered the phone and waited for a response from the other end. I was zapped! It felt like an electric charge. The hair on the back of my neck stood up, and the energy entered my chest and went out through my feet. It left me weak. All I could do was to sit on the closest chair. Before I had a chance to say ASSHOLE, I heard a click on the other end. Danny must have used the vibration of my voice to send the charge. I had thought about tracing the calls, but he could be calling from different locations. How long had he been planning this? Thirty-six years was a long time to wait for revenge. Did Danny think I would go back to him?

CHAPTER 7

Taking Walks

To expend the energy that was consuming me, I started taking longer walks. I enjoyed them most in the early evenings, after the day's work was done. I put on my sneakers, opened the front door, and stepped outside. An invisible weight immediately lifted from my shoulders.

The sunset in the west splattered and painted the sky with vibrant gold and pink cotton candy clouds. Palm fronds cast shadows on the canvas, with orchid and cypress trees contributing their different shapes. I breathed deeply, inhaling the cool, relaxing fresh air. *Thank you, God, for the oxygen.*

I waved to neighbors arriving home late from work. I sensed the aroma of chicken cooked on a patio grill. I inhaled the fragrance of gardenias and jasmine. I felt the essence of Spirit.

Neighborhood children playing ball in the street paused to let me walk past. I enjoyed the mockingbirds as they entertained me with a medley of song. I walked through the neighborhood where the grass met the street on my habitual path. Warren, who lived on the corner, was sitting out outside and waved to me.

"Carol, I have an extra chair here. Would you like to sit for a while?"

LOVE FROM THE OTHER SIDE

I gave a friendly wave. "No thank you." I wanted to keep my steps in rhythm. I wanted to walk. I patterned my route and circled back home. My legs felt heavy. My mind was clear. I sensed a healing. I was relaxed. I'd sleep soundly tonight. Losing weight happened to be a nice benefit. I certainly needed to do that.

After Danny's hang-up calls, I started to meditate and say my rosary every morning. Spirits who stay earthbound need help to cross over. I wanted balance and closure. Carrying a cup of coffee from my kitchen to our bedroom, I sat on the bed in my blue cotton robe and faced the rising sun from my window. My connection with Spirit felt peaceful, and helped me to survive the coming day. I had to admit that my ego and feminine pride had not been lost. It felt good to think, after all these years, that someone still thought about me. I was fascinated and also troubled by the mystery and romance of Danny's energy. I was trying to turn what I thought might be negative energy into a positive. I was trying to remember the message of my dream while I was in the light. What was my lesson here?

Alex had been losing weight steadily for the last two years. I felt his tension and stress. He was worried about the bills. As a result, he drank more and ate less. We were not able to get medical help. We had little money and no health insurance. He was too young to collect Social Security. We were always a month behind on our rent. The money Alex made from tile contracts covered the cost of materials. It paid workers, and also covered gas and truck repairs and food for our table. But then we had to pay for city and county licenses, and taxes. Alex and the guys did excellent work. But the advertising we could afford was limited, and larger companies consistently under-bid us.

Often, in the mornings, I turned to Alex to feel his energy and warmth. He would say, "*I love you.*" His Old Spice deodorant smelled so good as I wrapped my arms around him and laid my head on his chest. I would look him in the eyes and say, "More ---- I love you."

To which he would reply, "More."

Alex worked hard in the hot sun, covering pool decks and driveways. He rarely complained now, after living in Florida for so many years. My son Jerry and son-in-law Henry sometimes took time off their jobs to help him when he needed it.

I was alone. It was a late June morning. While standing by the kitchen table a faint aroma of cigarette smoke filled the air... *Danny's here.* The radio station was playing a Brian Adams song, "When a Man Loves a Woman." I could feel the energy stirring in my stomach. The serpent "Kundalini" energy spiraled through my center chakra and into my heartbeat. Thoughts swirled in my head. I could feel a sadness and the heaviness of spirit energy on my chest, oppressive, like when a loved one has died. I knew Danny's entity had entered our home. After placing a load of laundry in the washing machine, I watched a morning talk show on television about marriage and relationships. There was something in the air. I could feel it.

I had just set a glass of iced tea on the coffee table when I heard a voice.

"*Do you know what it's like to live thirty years with someone you don't love?*"

The energy I felt startled me; I spilled my glass of tea all over the coffee table.

That was why he married so suddenly! I felt sure he had gotten her pregnant. When I heard her name, "Karen," I realized she had attended my high school, but I had barely known her.

"You made your choices!" I yelled.

"*And so did you.*"

The thought he sent came back to me. I knew I had a problem. I thought Danny looked so alive at the flea market. Something else was going on.

If it weren't for the meditation circles I attended and the spiritual knowledge I had received from my family, I would have been headed for a nervous breakdown. I could just see myself searching for a psychologist and staring at ink blots.

LOVE FROM THE OTHER SIDE

My mother had spoken often of her dreams and premonitions. There were many happenings in our family, some stranger than others. On the day my Aunt Ruth passed away, there was blood on her living room wall. Three tiny streams of a red sticky substance were under each plaque of Jesus and Mary that had been part of her decor for as long as I could remember.

Sensing spirit energy and reading New Age books is one thing, but having the actual experience with Danny was something else. I needed help with this. I could think of a couple of people I knew who might be able to shed a little light on the situation. I continued my household chores, deciding that one Sunday soon I would visit a local spiritualist and mystic, Reverend Edward (Red) Duke, and attend his evening service. I didn't have much money, but I knew he would do billet readings for a small donation. I had not seen or talked to Red for several years, but I had known him for three decades. Other friends I knew had received help from Red and spoke of him with reverence.

In the meantime, the other person I could contact was a spiritual counselor and astrologer who just happened to be my nephew. Mark might have some helpful insight and suggestions.

Of course not all of my relatives were likely to be helpful, Alex just put up with my psychic experiences. My son Jerry jokingly called my friends Fruit Loops and thought that anyone who believed in "that spirit stuff" had to be a little nuts. I would remind him that he didn't think that way when his grandma was alive and kidded him that with his lifestyle he should watch out, the spirits would get him. He was just embarrassed to admit that his mom could feel spirits, and preferred that I not mention it to his friends.

CHAPTER 8

Seeking Help

I realized that I needed help to understand why Danny's spirit was in my life. My friend and nephew, Mark Dodich, had knowledge and experience in spiritual fields, being a well-known astrologer and spiritual counselor. It had been almost a year earlier that he moved from Florida to Oregon, but we still kept in touch. I wrote Mark a letter explaining what had been happening as best I could in hopes that he could help shed some light. Why did Danny's spirit choose now to enter my life? How could he materialize and also find the energy to use our telephone? I needed to find answers.

While he had lived in Florida, we would occasionally drive around and check out the metaphysical stores in Fort Lauderdale, Hollywood and Deerfield Beach together. Close to six feet tall with dark brown hair and brown eyes, Mark would feel his way around the stores. Usually, he would be drawn to crystals or books by the attraction of the energy he felt. Sometimes a book would just fall off the shelf near him, a definite attention-getter.

One of Mark's working methods would be to go into a trance, and Spirit would speak through him, a technique frequently referred to as channeling. While channeling for us on one occasion, the spirit being channeled identified itself as Osiris, an Egyptian symbol of the god of creation. On a previous occasion the spirit being channeled identified itself as Saint Germaine, who it has been claimed was reincarnated as Sir Francis Bacon.

With his background and credentials I thought him qualified to help me with my dilemma. He studied social science, and graduated from Kent State University. Since 1980, he had provided astrology and intuitive consultations to those who asked for the insight. He is a certified astrological professional from the International Society of Astrology Research. Mark told me his purpose is to serve the highest good by allowing the light and love of our Divine Source to flow through him. I was really hoping for some knowledge or answers.

Within a week, I received a letter to my question.

Hello Carol,

You asked about all of the hang-up calls you've been getting. There could be a lot of reasons -- both metaphysical and physical. When I worked for the phone company in Texas, we had a lot of problems with the switching equipment when the weather was especially wet or humid, which resulted in "ghost calls."

On a more metaphysical level, it could be that the change in energy of anyone in the house could be creating some electromagnetic disturbances. If you are living in harmony with your inner being, it is simply that you are increasing your ability to channel the increasing frequencies of energy on the planet. A book I read recently, called Awakening to Zero Point, discusses that the magnetic resonance of the planet has decreased more than 30 percent over the past 2000 years. About six years ago, the electrical frequency of the earth rose from 7.8 to 8.6 hertz. This means that there is a reduced buffer zone of protection from our thoughts.

This is good to the extent that our thoughts are of light, as they manifest the love vibration and our needs faster. However, thoughts not in harmony with increasing frequency smack back harder since there is not a dense vibration to reduce the impact.

My personal feeling is that those not in harmony with changes that have begun will, on some level, want to leave this earthly vibration for a place that is more willing to wait for change. So, a change in frequency of anyone in the house could be playing with the electromagnetic fields of the phone or other electrical appliances. At least, it's something to think about.

Blessings, Mark

I was pleased that Mark had taken the time to give me a detailed answer. But it didn't solve my problem. I chose not to show my letter to Alex. I wanted more proof about what was happening to me and in our home. I knew Alex wouldn't believe me.

On Monday, July 11, Alex walked in the front door about four o' clock, covered in cement dust. His blond hair was gray and he carried a six-pack of Budweiser. I glanced at him as I stood by the kitchen sink, and then I added water to the last of the liquid dish soap in the bottle. I always tried to get the most value for my money. I had just washed breakfast and lunch dishes. I could see that Alex didn't feel well. He put five cans in the fridge, and kept one can to drink. He sat on a blue and white vinyl-covered chair at the table and popped the seal. Princess carried her toy over to Alex and plopped it in his lap. She sat by his chair and gazed at him hopefully with her large brown eyes.

"Not right now, Princess – it's too hot outside." She lay on the tile floor beside him.

"You look tired. Have a hard day?"

"We'd just finished laying out the pool deck when the sprinklers came on and washed the material off. We have to redo it tomorrow."

LOVE FROM THE OTHER SIDE

They were working on a resurfacing project. A trowel is used to lay a thin coat of cement on existing concrete. Then a design is cut in to simulate stone or brick. On the second day, the pattern is grouted, colored, and sealed with water-resistant material that penetrates through the foundation. Lifting the fifty-pound bags of material and cans of sealer was hard on Alex.

"How much material did you lose?'

"About three bags -- if it rains tomorrow, we'll be behind schedule."

"Did you eat lunch?"

"You know I can't eat in this heat."

"But every day I prepare a sandwich and fruit for your lunch, and then you come home and start with a six-pack on an empty stomach. Why do I bother making your lunch? Why don't you take better care of yourself?" I didn't want to harass him, but I needed to say what I felt.

I was worried about Alex. Not long ago, I had had a dream. The energy felt like Alex, and his funeral. The color was red, and the action erratic. I awoke with a heavy heart. I was disturbed. I know my premonitions, but I know I can't always change the outcome. My fear was Alex would be so depressed he would end his life.

"Quit nagging me!" He finished the last of his beer and opened the fridge for another. He sighed deeply, and looked at me.

Dear God, I know, somewhere deep down inside of me, that Alex is not going to stop drinking. I do not want to argue. Miracles are few and far between. I just want to keep the peace.

"You know, Carol, customers I worked for, so many of them lately, have been married for years. Now they are separating. This lady I'm working for now -- she's a short Jewish lady, reminds me of Dr. Ruth. She told me, 'My husband left me once for a skirt. He came begging me to take him back. He found out she had the same thing under her skirt that I had under mine. I gave him a hard time but he's back.' I laughed with her. She made my day.

SEEKING HELP

The stories my customers tell me are overwhelming. I'm going outside to organize the back of my work truck, while you start dinner."

"Don't be long." I watched Alex pat Princess on the head.

"Yeah, sure." He threw me a kiss. "I love you."

"More. By the way, it has been a few years since I visited my psychic friend Red. I'd like to attend his evening meditation service this coming Sunday. Come with me. It will help you relax."

"I'll think about it. You know I don't believe in that spirit stuff, but I haven't taken you out for a long time. If it will make you happy, we will go together."

He put his empty can in the trash and headed for the front door. I put a pot on the stove for spaghetti.

After dinner, I saw Alex with a beer in one hand and a Frisbee in the other. He opened the front door of our small two-bedroom one-bath home. Princess followed him out the door. Alex sat in one of the lawn chairs by the front window. He would throw the Frisbee and Princess would run happily and catch the flying round disk. She never missed. I heard tires on the gravel in the driveway. Through the front window, I watched Alex's helper Ray climb out of his truck. He waved at my reflection in the window. I waved back. Alex's friends stopped by easily; they knew he would help. I looked at his cut-off jeans and white tank top. Ray was short, maybe five feet two inches. He had dark short hair and his white sneakers looked new.

"Hi boss," I heard Ray say to Alex.

He sat down in a lawn chair. Princess greeted Ray with the Frisbee in her jaws. Ray stroked our Shepherd with his hand and threw the Frisbee. "We were held up on the job today -- I was hoping you would front me some cash until tomorrow."

Like us, Ray barely made it from paycheck to paycheck. I stepped outside and offered him a beer. Alex handed Ray a twenty. I knew it was probably his last twenty, but that was Alex.

Ray sat for a while. "Hey boss, I was wondering if we could

get our old bowling team back together. I keep thinking about the night you bowled that 300 game. Everyone at the alley was hollering and patting you on the back. We had some good times."

Alex drank a swig of his beer. "Yes, we sure did." He smiled. "I can't see it right now until we're better off financially, Ray. I can't make any commitments right now."

Ray stayed for a while, finished his beer, and played with Princess. When he stood up to leave, he said, "Thanks for the twenty, boss. I'll see you in the morning."

I looked out of the front window and saw Princess lying down in the grass. She looked at Alex with her large brown eyes. They entered the house and both sat on the worn light-green sofa to watch TV. Though our Shepherd topped ninety-eight pounds, Princess sometimes thought she was a lap dog and jumped up to keep Alex company on the sofa. Eventually she rested comfortably, with her chin on his knee. They sat together in a familiar pose.

Later that evening, after my shower, I lay in bed feeling melancholy. I heard Alex singing in the bathroom as he turned off the shower. He walked in the bedroom and placed a screw on the nightstand.

"Oh my God, Alex." I smiled. "Are you all right?"

"You know it, baby. I'm here for you."

CHAPTER 9

The Haven for Spiritual Travelers

It wasn't too long before Alex agreed to attend the evening service of a local psychic with me. I was really hoping to receive information about Danny and his spirit visitations. On a Sunday evening, in July 1996, we dressed in jeans, T-shirts, and sneakers. We left in our blue Mazda work truck to visit "Red" Duke and The Haven for Spiritual Travelers. The service started at seven.

We talked a little on the way to The Haven; I was hoping to keep Alex in a positive mood. "I know you haven't felt well for a long time. There is a lot of good energy in this group meditation. I'm sure it will help you."

"Well, Carol, if there are miracles, I could use one."

"Alex, I know you're not into my psychic world but I am happy you're willing to share this evening with me. You've never met Red or some of my friends." I wanted to tell Alex about my hang-up calls and the day I was zapped, but I was afraid he wouldn't believe me. I needed verification. In time, I knew that Red could give me support. Watching Alex change lanes on I-95, it was plain that traffic was unusually heavy, and this was not the best time to mention it.

Alex drove into the front yard of The Haven for Spiritual

Travelers, and parked our truck. The home was easy to find. Red's thirty-foot flagpole was always lit with white lights, and his American flag waving in the wind. Reverend Duke directed cars into parking spots on the empty lot beside his house, cheerfully greeting friends as they walked toward his home.

Red could be called a down-home country boy, with simple, practical common-sense ways. He was in his late sixties, and stood a little over six feet tall with a slender frame. His red hair had no gray at all and was tied back into a ponytail. The Haven for Spiritual Travelers was Red's habitat. It was an old house built in the 1940s, with a white stucco foundation, a brown tile roof, and wood floors. His home was in good condition, for being built over fifty years ago.

I always felt comfortable at Red's. We stepped into the front Florida room on the familiar green and white tile floor, for the evening service. I listened to meditation music as it played softly through the speakers. The aroma of frankincense incense drifted through the room. Blue vinyl chairs were lined up in rows of four across. Alex chose two seats in the back of the room. I reached for one of Red's pamphlets on a desk in the corner, and looked at Alex.

"I want to read this to you, because you are not familiar with the way Red does his readings."

Alex rolled his eyes. "I'm your captivated audience."

"Pacify me, please."

I read from his pamphlet. " *'Red feels tremendous compassion for other people's sufferings. He learned this as a result of medical problems he has had since serving aboard ship in the Navy, and surviving atomic radiation poisoning at Bikini Atoll in the Marshall Islands in 1946.'*"

"Wait a minute -- you're telling me Red survived radiation poisoning?"

"Yes, Alex. Remember, you were never interested in my friends. I went with you on your bowling nights, but now it's your turn to show me some support. Now, let me continue reading. The next part is how Red communicates with Spirit. *"If there is something you do not understand, please don't hesitate to ask a question. Red could be channeling a spirit who speaks in a foreign language. Please recognize that Red can go rapidly from one dimension to another.'"*

Alex was looking at me with his head cocked and eyebrows up. "No way, Carol," he said. "Who is Red, Jesus or something?"

"Alex, quit. Now, pay attention. The man is gifted, he's helped many people. Let me finish."

I continued reading softly, " *'Red believes in simplicity. The power of Spirit draws energy through him. All of this is natural. Nothing negative is permitted in The Haven for Spiritual Travelers, or in any teachings of spiritualism and mysticism. At times, Red will feel a psychic impression around you. He must ask your permission before he channels a healing.*

" *'When Red is channeling a reading, please answer him. If you do not understand the message, say, "Please clarify that." Voice vibrations open up additional passages of information that are important.'"*

I saw a pile of billet envelopes on a table by the front door and recognized them as donation envelopes like those used in my Catholic church. I looked at Alex.

"Do you want to ask a question?"

"No. I don't need Red or anyone else telling me my future. I know where I'm going."

"Red is very good at what he does. You could just appease me."

"Carol, I don't want to ask any question. I'll just watch. It should be interesting."

"Well, keep your eyes and ears open. Red's Haven for Spiritual Travelers is haunted. You could feel cold energy near you or see a shadow pass through his wall."

"Now I'm worried." He laughed; at least he seemed relaxed.

LOVE FROM THE OTHER SIDE

Dawn, Red's office assistant, stepped toward us. She brushed a strand of dark hair off her shoulder and spoke directly to Alex. "I overheard you talking, and I want to tell you, I have slept over when Red had to leave town for a few days. He has a bass fish that sings when batteries are in it. Well, around four in the morning I was in a deep sleep when I suddenly sat up in bed because I heard someone singing. I got up and started searching. When I got to the parlor I found this bass singing *'Don't worry, be happy.'* It was when I tried to turn it off that I discovered there were no batteries in the fish."

The look on his face was entertaining. I smiled at Dawn and said, "Thanks for confirming your support in the reinforcement of our spiritual world, Dawn. Alex thinks we're all nuts. I came back because I need answers."

"It's nice to see you, Carol. What kept you away for so long?"

"I've been busy with the family and helping out with our tile business…phone calls, and scheduling appointments. But tonight I'm here on a mission; I hope Red can help me."

"I know Red keeps billet questions private. If anyone can help you, I know Red can." Dawn smiled at us and walked on toward some other people.

"What do you need help with?" Alex asked me.

"I told you I saw my old flame at the flea market a while back. And since then I've been getting a lot of odd hang-up phone calls."

"Take me to him. I'll kick his ass."

"Alex, I think he is dead."

There was a pause before he replied thoughtfully to that, "You need help."

I sat down beside Alex and followed instructions on the envelope. I wrote my first name for a feeling of the past. Second, I wrote my middle name for a feeling of the present, this is emotion. Third, I wrote my last name for a feeling of the future, this is Spirit.

THE HAVEN FOR SPIRITUAL TRAVELERS

On the second line, I printed my first, middle and last names. I placed a few dollars for a donation in the envelope and wrote my question.

I feel so good when I come to your home. Thank you for being here, Red. Can you tell me anything about my hang-up caller? I walked up and placed my billet in the box by the podium.

People gathered in the room, which held twenty comfortably. Red took his place in front of the podium. He was dressed in casual brown slacks and a plaid sport shirt. Light reflected on the Kokopelli charm that hung from a chain around his neck. Kokopelli is a Hopi Indian in spirit. He plays the flute and is considered a wandering minstrel. He carries a sack on his back that is filled with seeds and rainbows, gifts from Spirit. Red's mother was a full-blooded Cherokee Indian. It is how Red came to use Shaman energy for his healings. His father was Irish, which probably had a lot to do with his personality, not to mention the red hair.

"Good evening everyone." Red had a warm smile and deep, commanding soft voice.

"Good evening, Red," the group responded.

He smiled and warmed up with his usual jokes. "I grew up in the South and we were all a bunch of rednecks -- still are, if you know what I mean. One day I heard my dad ask our neighbor, 'How many children do you have?'

"'Nine boys,' she said.

"I heard my daddy say, 'Really? And what are their names?'

"'Henry, Henry, Henry and, well, they're all named Henry. That way when I wants them to come in from the yard, I just yells Henry! And when I wants them all to come to dinner, I just yells Henry!'

"'What if you just want to call one of them to do something?'

"'Then I calls him by his last name.'"

Everyone chuckled.

Red liked to talk about his Navy days. Previously, I had noticed a sign on the wall by the door. It read: *It required Arthur Mole* and 10,000 male sailors to make a flag in World War One. It took but one female to make a Wave in World War Two. (William Phinney.)

The group sang a few songs to get the energy and vibrations working, like "Just a Closer Walk With Thee." One song was "I Have a Dream" by Abba, which is a personal favorite. I am always fond of the phrase, "*I believe in angels/something good in everything I see.*"

Red walked down the aisle saying hellos, shaking hands, and patting friends on the back. His hands were large and warm as he held mine. The energy felt good.

"Welcome back," he said to me.

There was a message printed on his monthly calendar. *Good teachers lead us toward the light. Great teachers are our light.* After his lecture, he read a verse from the Bible. We had a group meditation and he read a benediction. The group dispersed for a ten-minute break before the billet readings began.

Many friendly souls attended the service. I always enjoyed meeting and talking with them. Alex had stepped outside the front door to enjoy a cigarette with other smokers. I retreated to the backyard to walk on the flagstone paths among the ferns, palm trees, and jasmine.

It wasn't long before I heard Red ringing his cowbell from the Florida room. Spiritual friends started returning to their seats. He made announcements about classes. Then he began to read billets.

"I will not read the questions on your billets out loud. I respect your privacy. If you need to talk to me after our service, please let me know."

Red picked an envelope from his box and slowly ran his hand over it. Then he called the name Sue. A young woman seated by the wall raised her hand.

"Yes Red, I'm Sue."

He looked at Sue and asked, "Let me come into this vibration."

"Yes."

"Oh my God, I feel your pain. Do you suffer from heart problems?"

"Yes, Red."

"Can I work with you for a minute?"

"Please, Red." Sue stood up and I saw how her color was ashen gray. She looked to be in her late twenties. The crowd watched as Red from about ten feet away raised his hand and made a waving motion.

"Doesn't that energy feel good?" Red smiled warmly toward Sue. At that point I saw the young lady's skin start to turn from gray to red, then to pink. Within about three minutes her skin was a normal color. When he finished, Sue walked to the front and gave Red a hug,

Wiping tears from her eyes with a tissue she said, "Thank you, Red," and between tears, "My pain is gone and I feel so much better."

"We believe in simplicity, there are no tricks or gimmicks here."

Red reached his hand in the box and picked up an envelope.

I knew the billet was mine. My heart was racing but I tried to stay calm. I raised my hand. "Hello, Reverend Duke." He insisted people speak up.

He looked surprised and then smiled. "Nice to see you again, Carol. We missed you. Please call me Red. There are no stuffed shirts in my dominion."

He smiled while he ran his hand over my billet. He looked around the room.

"We are called The Haven for Spiritual Travelers. Friends are welcome to come and go as they need us. We are here to help and to teach. Many souls have traveled through this house, past and present. Let me come into this vibration?"

"Yes, Red," I said.

"This person just wants to make some noise. Rattle a few

chains. You will get through this and be a stronger person for it."

I felt a bit unsure and confused; I'm sure I looked upset. "Thank you, Red. But I was hoping for more support."

"God won't give you any more than you can handle. You know I'm here if you need me."

I gulped, and thought, *Having trouble with an old flame is one thing. But the psychic attack I had, the hang-up phone calls and what I thought was negative energy around our home is something else. How far and for how long is this going to go on?* It was a wait-and-see situation.

"Can I read part of your message out loud?"

"Yes," I said. I knew he would never do anything to embarrass me.

"Your home gives me a feeling of peace. Thank you for being here, Red." After reading my written statement he looked at me and smiled warmly.

"Thank you, Betty," he said to me. Then to the spiritual friends, "You know I have to get my strokes in where I can."

"Red, it's Carol. You called me Betty?"

"The name Betty just keeps coming up."

"Okay, Red. I'll let you know when I meet Betty."

It didn't bother me that Red called me Betty. He was always pulling names out of the air. I figured there was a Betty somewhere in my future.

When the billet readings were over, the crowd disbanded for coffee, tea, and snacks. Alex stepped outside for another cigarette, so I went to socialize. In the center of the parlor sat a pot-bellied stove. A multi-colored circular braided rug rested on the wood floor in front of it. People sat on a well-worn sofa along the south wall covered with a striped sheet. Several people admired the antique hutch filled with ceramic statues of angels, family photos, and ceremonial crystals. A round table in the corner held large trays of homemade brownies, and I treated myself to one. Then I sat in the wooden rocking chair by the parlor door.

A friend I had known who referred me to The Haven several years ago came over and stood beside me.

"Hi, Carol; I haven't seen you for a long time. How have you been?"

"Hi, May. I like your blond hair. It makes you look younger."

"Well, I just had it done. So tell me -- what have you been up to?"

"Well, May, I'm here tonight because I'm being stalked by an earthbound spirit. I was hoping Red would help me."

"Well if anyone can help you, it's Red. You know when my little girl drowned in our swimming pool it was a horribly traumatic experience. Her spirit was active in our home. She was hiding my car keys, and we had all kinds of happenings. If it wasn't for Red I don't know what I would have done. He helped me find peace."

"May, I know you've told me some stories about your little Lucy. I'm so happy you have found peace. Call me sometime soon so we can talk."

"I'll do that, Carol, and you take care. Red will help you, I know it."

The aroma of fresh popcorn drifted from the kitchen. I joined a small group I overheard discussing positive energy and a book entitled *The Celestine Prophecy*. Tomorrow morning, I told myself, I would buy it.

We said our goodbyes. I exchanged hugs with Red and with a few friends there. Growing up in my family, hugs were few and far between. Being embraced by friends helped me to feel accepted and welcome.

On the way home, I asked Alex, "Were you able to relax tonight? Did you enjoy the service?"

"Carol, I came because you asked me, but this is not for me. I was surprised at some of Red's messages. People showed real emotion. But give me a six-pack and a football game. Then I'm relaxed. And I didn't see any ghosts."

"I can't force you to understand my spiritual experiences. If it just doesn't interest you, I won't ask you to come again. I guess you're not comfortable with the service."

"Did you receive the information you wanted to hear from Red?"

"No, I have to admit I'm disappointed. But I'll keep coming back to The Haven until I receive the answers I'm looking for. What I want is closure."

"Carol, I have to admit you're persistent."

"If you could see and feel what I'm going through, you would be looking for answers too."

We were quiet the rest of the way home. I knew I wouldn't call on Red unless it was necessary. He was very busy and actively involved with people who needed help or healings. Somehow, I was going to get through this.

When we arrived home, I threw my purse on our bed. Alex headed for the sofa and turned on the TV. With the dog beside him, he popped the seal on a can of beer. In the kitchen, I filled a glass with iced tea, pulled a chair out from the kitchen table, and sat down. I reached for the telephone and dialed my sister Emily's number. She was the middle child and ten years older than I. She still lived in Ohio where we were born, and nourished roots I had left in my hometown. Emily kept me informed about family, friends, and neighbors.

She answered after the third ring. "Hello."

"Hi, Sis. I just need to talk. Are you busy?"

"No, I'm just sitting here in the kitchen going through old mail. Are the children okay?"

We talked for a few minutes about the usual family stuff until I got around to telling her about the day at the flea market when I saw Danny and the energy shock when we connected. When I got to the hang-up calls and the unease I had been having since, her response was serious concern.

"You're kidding!" she exclaimed. "How did he find you?"

"I have no idea. I haven't heard from Danny or Irene since

Alex and I married in 1962. I'm not sure what he wants after all these years. Would you ask around? By the way, I was sure you told me a few years ago that Danny had died. Please, see what you can find out for me."

"I don't remember telling you that, Carol, but I'll do what I can. I'll call you when I know something."

"Thank you. I need to go now. Love you. Bye."

As I hung up the phone, I thought about Danny's sister, Irene. I hadn't heard from her since I broke my engagement. After our big fight, I was hesitant about talking to anyone in Danny's family. Irene never tried to contact me. What was going on? Why the psychic attack?

After talking to Emily I was restless and paced the kitchen floor. Suddenly, I now had the urge to write, and started to pen my thoughts. When I started writing, the action would not leave me alone. I paced the floor night and day. I made notes on little pieces of paper or tablets. I wrote about my mother, my family, and the experiences I had with Danny. Often a thought would come to me and I would write it down. I kept a detailed record of my dreams and dates. Developing a sense of urgency, I knew that writing was what I had to do. It was time to speak out about my private life with spirits. There was no way I could just push it down inside of me and keep all that was happening to me. I don't like hiding in a closet; I get claustrophobic. Communication was my only outlet to peace, healing, and understanding.

CHAPTER 10

Spirits

The next morning, I drove to the local metaphysical bookstore. I found and bought a copy of *The Celestine Prophecy*. I overheard the salesperson say, "If a spirit speaks, it weakens their energy." Okay, I thought. *The saleslady was giving information to another customer; the coincidence was that it helped me.* A light flashed in my brain.

This may have been one of the reasons why Danny, my hang-up caller, was not verbally communicating with me. He was saving his energy.

It had been a long, busy day and I was ready to enjoy a brisk evening walk. Alex was sitting in the front yard playing with Princess. I leaned over and gave him a kiss, then left the yard.

Princess watched me leave the yard and Alex shouted, "We'll be here when you get back."

After my walk I watched TV with Alex and before I knew it, it was midnight. I was extremely tired as I stepped into the shower and felt the relaxing warm water flow down the back of my neck.

That night in the bedroom, Alex was sound asleep. Princess slept by the foot of our bed. I slipped into my comfortable pink cotton nightgown. Then I lit incense to help reduce the odor of beer that surrounded Alex. I chose a meditation tape, "Chant," by the Benedictine monks. Hitting the play button, I slipped into bed. Something made me uneasy. I felt a ping in the pit of my stomach. I waited uncertainly.

I was finally relaxing when I felt energy around my feet and legs. It was like when you're outside and slip your shoes off -- your bare feet feel a cool breeze flow across them.

Oh God, I thought. I had watched a program about spirit seduction on "The Other Side With John Edwards" and I read a book about sex with spirits, but I couldn't believe it was happening to me. Again I wondered whether this was something to do with my prophetic dream. The energy became stronger. I felt determination from the entity. I couldn't believe, as I lay next to Alex, that an entity was in bed with us. It was Danny's spirit and I wanted to communicate.

I sent a thought. "You know with this special gift you have and the strength of your energy, you should use it to help and heal people. Danny, you are in my past and I'm married to Alex now. You need to move on, go into the light -- you will find peace there."

I felt a brief moment of surprise and hesitation. Then it coiled its way up to my root chakra, like Kundalini energy. *Okay, I thought, I'm going to deal with this.* When I was lifted to the light, I knew I would be all right. As a warm feeling began to play with my sexual urges, I sent another thought.

"You know I was never easy, so give it all you've got." I was shocked and surprised. I didn't know what else to do. The energy moved up my spine through my chakras. It tickled a nerve in the small of my back. This caused a spontaneous reaction. I arched my back and took a deep breath. The entity gained strength. I heard the relaxing chanting of the Benedictine Monks, humming low on the tape, "Gloria in Excelsis Deo." Feeling helpless, I sent a thought.

"*Can you give me back my youth?*"

"No."

I was surprised to hear him respond.

"Can we travel?"

"No."

"You looked so good at the flea market. Seeing you stirred old feelings I had forgotten about. But I'm married and have a family -- why are you here?"

Danny did not respond. The energy continued into my heartbeat and stayed with me all night long. I thought about my past, and my marriage with Alex, the children growing up, and the different jobs I had. I worked in a box factory where my arms were cut from stripping boxes as they came off the press. Calling numbers in the bingo hall, paying winners their prize money. I remembered when Janet was in the hospital with diabetes.

There were no more responses or words from the entity, Danny, but I knew he listened to my thoughts. I reached a point of satisfaction. It was nothing like the orgasm reached by two people making love. Nevertheless, I knew that I had yielded to the dominant spirit of Danny's entity, leaving me with the reality of surrender. My mortal and spiritual worlds collided. I did not feel comfortable with spirit in our marriage bed. Alex was mortal and a spirit was a spirit. I didn't refuse Danny's spirit because of my experience when I was in the light. But I did feel unfaithful to Alex. Alex began to stir. I turned toward him and placed my arm around his warm body and held him close.

Alex awoke at his usual five a.m. I lay there trying to relax. My mind was racing. I reached for the pen and paper that I kept on my nightstand to record my dreams when I wake up in the mornings. At times, I have used automatic handwriting, when Spirit speaks through me. I started to write. The energy was forceful and the message clear.

"Do you want revenge because I left you?" I wrote.

"Yes! I love you," his spirit wrote through me.

This was not going away. I couldn't just ignore it! I believed my message -- the one I couldn't remember while I was in the light -- was about Danny. Red told me I would get through this and be stronger for it. *I will*, I thought.

When Spirit's light came through me, and the energy was right, messages I received in automatic handwriting were fantastic. Sometimes, spirits writing was like chicken scratchings, and I had to decipher the message. Also, the words would be connected. The i's are not dotted and the t's are not crossed. I put the pen and paper down, jumped out of bed, and put on my robe. Why was this happening now, after all these years?

I felt manipulated and controlled. I believed in the sanctity of our marriage. I was not going to let a spirit come between us. He knew my weakness. My husband and I were hardly ever having sex. After last night, I was tired and had no energy. The aroma of coffee drifted in from the kitchen.

I swallowed the first sip of coffee from the cup Alex had poured for me. Bacon and eggs were in the frying pan. I put two plates on the table. Alex liked to cook breakfast.

"You look tired this morning -- there're dark circles under your eyes," Alex said.

The telephone rang. He answered it. "Must have had a wrong number; they hung up."

"I told you we have been getting hang-up calls. I didn't sleep well last night." I was agitated. I picked up my coffee cup. "Thanks for cooking breakfast, but I can't eat right now."

"Are you all right?" Alex asked, the concern evident in his voice.

"One day when you're ready to listen, we will talk. I'll eat later, after my stomach settles." I returned to our bedroom, said my rosary, and meditated. I was leading two lives ... I could not move on until I found closure. Alex was easygoing, and we rarely argued. I did not like conflict.

CHAPTER **11**

The Supermarket

Alex walked in the front door that evening. He looked exhausted and was covered with cement dust. But he was singing, *"Un-for-gettable, that's what you are/ Never before has someone been more un-for-gettable."* I could hear the tiredness in his voice, but he was actually smiling. He put a twelve-pack of beer in the fridge and told me to make a grocery list.

"We're going shopping tomorrow." Hugging me, he said, "I just signed a nice-sized job."

"Oh, thank you God, for our abundance." I felt his sweaty arms around me and breathed in a manly smell that belonged only to Alex, my hands on his damp T-shirt. Our pantry was nearly empty. I looked at the two cans of beans and one can of tuna fish on the open shelf.

The following day we went to the local supermarket. I stood in front of the deli counter, waiting to order cheese and lunch meat. A young girl in her early twenties walked up and stood beside me. She stood too close to me invading my space, and seemed intimidating, with her stocky build. She stood about six feet tall. Her short curly hair had red and blue dye in it. Her arms were folded. She just stood there, looked odd, and stared at me. Just then, Alex walked up behind me. Alex put his hands on my waist and tenderly kissed me on the neck. He made my knees weak, but I knew it wasn't Alex. I sensed Danny had walked into Alex. My husband and I did not share personal intimacy in public. The energy felt different. Okay, I figured Danny's spirit had entered the girl with red and blue hair and waited for Alex to walk up behind me. Danny had the gift.

After that, the presence was not only around the house, but it was inside of me. It sapped my energy. I felt butterflies in my stomach most of the time and they tickled my heart. I kept taking long walks in the evening and doing sit-ups before getting ready for bed. While I was on the floor, Princess would sniff my hair or lick my face with her wet tongue and I would break out in laughter. I was losing weight. My clothes were falling off me. I tried to talk to Alex. He did not want to hear about the spirit.

"When you're dead, you're dead," he said.

One morning, I was in the bathroom, brushing my long brown hair. I looked in the mirror at the crows' feet around my eyes. *Need to start using some sort of moisturizing cream.* Then I heard the name *"Guinevere"* in my mind. Now I'm not gullible, I knew the movie theater had been advertising Sir Lancelot and King Arthur, but was Danny playing head games with me?

The energy in our home became erratic. Alex was drinking more beer in the evenings. He was becoming an alcoholic and couldn't even see what was happening to him. I could sense a presence standing in front of him. Our German Shepherd would sit, look at an empty corner, and whine. Then trouble started and our good days became stressful. Our truck broke down. Jerry and Henry had to replace the engine.

As hard as I tried to keep the peace so we could all work together, things fell apart. After replacing the engine in our truck, a tire blew out and a ball joint broke. Customers called and tile contracts were voided. Money was short. Eventually our phone was turned off. We all contracted the three-day flu and our waterbed sprang a

THE SUPERMARKET

leak. I couldn't prove all of this negative energy was triggered by the spirit, but I knew spirits feed off of negative energy.

I just shouted, "GIVE ME A REASON TO WAKE UP IN THE MORNING, YOU S. O. B.!!!"

Six days later, the phone was ringing as I arrived home from an evening at Red's. Hearing it ring, I figured it was my sister, Emily. Sunday evenings at 11 p.m. was our quiet time for communicating. I picked up the receiver.

"Hello, Carol -- what's been happening? Did you see Danny again?" she asked.

"It's been crazy around here, and Danny's spirit is everywhere. I just returned from Red's. I went by myself tonight. The Dolphins were playing and the guys couldn't leave the TV set. Pre-season game, I guess."

"Did he have anything of interest to tell you?" she asked.

"He said, 'Love law, law love.'"

"I asked him to explain. He used a metaphor and told me there was a white picket fence keeping me in, and Danny out. There was a good crowd tonight; I hugged him before I left. He is the only one I can talk to who knows what is happening to me. Even if we don't say much, I know I'm not crazy."

"Well, I found some information. Danny died of a heart attack in 1983. He was playing poker with friends at the time," Emily said.

I felt a tug in my heart. *Love law, law love*, I thought. "But it's now 1996 and he looked so alive at the flea market! I've been getting hang-up calls. It explains what has been going on around here. The energy is so powerful."

"You must have had some sort of vision from the past. It couldn't actually have been him, he's dead," Emily said. "Maybe repressed regrets?"

I didn't want to press the issue. Emily didn't know everything that had happened, but she had given me the confirmation I needed. When Emily made a statement, there was no debate. I did not want to try to explain why I knew she was wrong. Yes, Danny was in my past, but that wasn't a one-time zap. There was still more going on here. We said our goodbyes, and I hung up the phone. *Dear God, do I need this? What was my lesson here? What does he need from me?* A wave of anguish fell over me.

The football game was over. Henry, Jerry, Ron, and their friends had left. I started to empty ashtrays and gather beer cans, glasses, and potato chip bowls. Alex did not want to hear about my spirit ventures. He was the one person I'd shared my life with, and I could not talk to him.

When the opportunity arose, I would have a heart-to-heart with my daughter. I needed to get a lot off my chest. Janet was manager in the shoe store where she worked and was always on call. I knew she usually had Tuesdays off, and would find time to share an afternoon with me if I asked her.

I felt blessed because our family was close. We all lived within a few miles of each other and were able to support each other. But we were all so busy with our lives; we didn't share enough family time.

CHAPTER **12**

Janet

In August, we were still getting hang-up calls. I wondered out loud if they were all ghost calls. Alex was impossible. "Ghost calls," he smirked.

"Maybe they aren't all ghost calls," I told Alex. "But why do the calls happen when I'm home alone? And why did the calls start on the morning after I saw his spirit materialize at the flea market? And why are you so closed-minded? My God, Alex, there is a whole universe of energy around us."

"Carol, I love you no matter how crazy you are."

Dear God, I refused be angry or upset. I phoned Janet and said we needed to talk. "Hi, Janet -- can you come over for lunch on Tuesday?"

"Sure Mom, I'd like that."

"You know, Aunt Emily mailed me a package with garlic in it. I doubt if the garlic will help me. Spirits are abundant."

"Be careful, Mom," Janet laughed. "Remember the Salem witch-hunt. Some people might not understand."

"Thanks for the warning, See you on Tuesday."

I reminded myself that our family was no stranger to spirits, or at least odd coincidences. On the morning my grandfather died, a sparrow sat on Aunt Fay's kitchen windowsill tapping on the glass, seeming to try to get her attention.

I had listened to Aunt Fay tell my mother, "I was fascinated by the sparrow's visit. It was at that moment my phone rang and I was told the news Dad had a heart attack." Clearly the sparrow was trying to give Aunt Fay a message.

Janet stepped through our front door. Although she was a grown woman, the sunshine on her blond hair gave her a halo, and she looked like my little girl in her shorts and tank top. Princess was by the door, ready to greet her.

"I love you, Princess," Janet bent over and ruffled her fur.

Janet poured two glasses of iced tea while I grilled cheese sandwiches. I placed two bowls of tomato soup on the table.

Where do I start my story?

"So what do you want to talk about, Mom?" She took a bite of her sandwich and looked at me with clear blue eyes.

"Remember Mother's Day at the flea market?"

"Yes," Janet said.

"Something happened to me that day."

"We had a good time, Mom."

"We did. But I saw an old flame of mine, Danny, from my high school days. I didn't say anything to you at the time. I was in such shock. Since then I've been getting hang-up phone calls. Janet, I am being haunted."

"Really, Mom; I'm surprised." Her eyes were wide with an expression of wonder. "But with all the visits you receive from spirits, how do you know it was Danny?"

"It's true, Janet, that I don't always identify a spirit. Consider that I saw him materialize at the flea market and then immediately had the psychic attack. It was too much and too intense to have been from a stranger. Danny's soul was in the light of his eyes. When we die our ego, etheric field, and astral shell all hold our personality. He appeared for a reason. I just have to figure what it was."

She looked surprised. "Mom, does Dad know about this?"

"I tried talking to your Dad. He just said, 'After thirty-five years?' He laughed. I guess Dad figured no one would return to a time and place after all those years. Remember the rash you had on your neck, when we first arrived at the flea market last May?"

"Yes. We thought it was heat rash."

"I believe you were given that rash on purpose. You know some mothers and daughters look alike. If Danny's spirit had planned this meeting, he wanted to be sure to send energy to me. So you were given a rash on your neck. Remember the first pog stand we stopped at? When I saw Danny, I had a psychic attack. The energy went right through me."

"I'm trying to remember, Mom. But I can't recall it." She set her glass of tea down and tilted her head.

"You know, when your friends call, they mistake my voice for yours. I have been zapped over the phone. So if you answer our phone and you're mistaken for me, you might get zapped. The energy won't hurt you. It's comparable to static electricity."

"Wow! So, Mom, how did you meet this guy? I want to hear about it."

"Well..." I took a deep breath. "It was in the mid-1950s. His family moved into a house down the street from ours. I was going to school with his sister, Irene. She and I went to school dances, double dated, and went to movies -- the usual kind of stuff teens do."

Janet glanced over to one side toward the dog, her eyebrows raised. "Mom, look at Princess."

I smiled. "You see, Janet, Princess is on her back with all four legs in the air. She is smiling like someone is rubbing her belly. Danny is here listening to us talk about him."

"Okay Danny, I want to hear about you. Mom, tell me the rest of the story, and Danny wants to hear."

"Well, Irene had mentioned her brother Danny was in the Army, stationed in Korea. I said I would write to him. He came home on leave and we dated. After six months, he was discharged from the Army and came home. We dated for almost three years and fell in love. He gave me an engagement ring on Christmas, 1958.

"After my graduation from high school, we planned a wedding. But Janet, at the time I started having second thoughts. I was afraid of being tied down to marriage when I wanted to do so many other things with my life. I felt a lot of pressure and didn't think I could handle it. Then one evening, I was walking alone down the street. I had a vision of him and a dark shadow. After that we had a fight because he lied to me about going to work. I gave his ring back and left for New York City."

"I remember. You have a picture of him in that shoebox you keep in your closet." Janet drained her iced tea glass. I refilled it, and she took another sip.

I received a message in thought from Danny. *I did what I had to do and I still love you.*

I ignored the message and continued. "When I married your dad, I burned all of Danny's old letters and pictures. I wanted to put that part of my past behind me. I did keep one photo of Danny and me at Christmas. Danny was dressed in a suit and tie. I had on high heels, and a black shift with red sleeves. We stood by the Christmas tree. You know I've been going to Red's, trying to find out the meaning of all that has happened to me. Red tells his spiritual friends, 'If you try to run from a problem, it will come back and smack you in the face.' He's always so right."

"Mom, do you think this Danny would hurt you?"

"If Danny wanted to, he's had plenty of chances. Of course, I can't be sure of anything. But no, I don't think he wants to physically hurt me. You know in 1959, I flew home from New York on Christmas Eve to be with Grandma and the family. I hardly had time to put my suitcase down when the phone rang. Danny's mother was on the phone hysterical, sobbing and begging me to come to her house. If I had, I would have given in to him. Instead, I wondered what he had gotten into and told his mother, 'I'm sorry, I just can't,' and hung up."

"Mom, are you saying I might have had black hair instead of blond?" We both smiled.

"No, dear, if that were the case, you would not be you. You know if Danny would have called me or come to the house, I just might have married him. I wasn't happy in New York. It was exciting working and being on my own, but I missed my family. I was lonely.

"Why did Danny's spirit choose now to come into my life, when your father hasn't been well for so long? Danny's spirit has been really playing games with me. I'm going to rely on automatic handwriting. I need to try and communicate, find out where his mind is. What will help bring closure to all of this?"

"Wish I could help you, Mom. You know, your clothes are baggy on you. You're looking good." She snapped the elastic on my waistband. "But you need clothes that fit."

"I hope to start working soon. Then I'll have money for clothes. On Mother's Day I was depressed. I haven't been happy for a long time. Now I've had a wake-up call. I'm fascinated by the mystery of it all. It was last Easter, when I had a dream and instructions from Spirit. It all began on Mother's Day at the flea market. I wonder if Grandma Marge had anything to do with Danny's visits. You know Grandma liked to be in charge of situations. I can still picture her now in the floral print cotton housedress she used to wear."

"Mom, look at the sunlight by the fridge."

It was a beautiful yellow light. There are dark spirit shadows and light shadows; this was like a shadow of spirit light. "It's not coming from the shade tree by the kitchen window. Your grandmother is here, Janet. She visits me often. I wonder what she tells Danny. I can see her short brown hair. Her round Polish face... she's smiling. I have to figure out my dream, and what Spirit was trying to tell me when I was lifted into the light."

"I remember when Grandma died. I smelled her perfume. I

know about your dreams, Mom. I have dreams too, you know, and I've inherited that part of your genes. I don't keep notes like you do, but those I remember do come true." She looked at me with her clear blue eyes. "Tell me more, Mom."

"Grandma's next door neighbor Edith called me the morning Grandma died. When I answered the telephone Edith said, 'You'd better come over here. Your dad needs help.' I wondered what my dad would need help with. Mother took care of everything.

"When I walked into their living room, it was filled with paramedics. Grandma had been in the shower getting ready to go to card club when she had a sudden heart attack. I watched Grandpa's grief through my own tears. He held a large white handkerchief and wiped his tears with trembling hands. I wondered if he could live without her."

"Mom, that is so sad." With a tear in her eye she said, "I remember when Grandma let my girlfriend Fran and me stay in her home for a sleepover. We had good times."

"You know, Janet, Grandpa never washed a shirt or a dish. I didn't think he could boil an egg. I stayed there that night. I didn't want to leave him alone. The spare bedroom was musty. But after making telephone calls and arrangements all day, when I finally lay in bed, I was too tired to get up and open a window. I closed my eyes but I couldn't sleep. I kept wondering what would happen to him. My eyes filled with tears when suddenly I sensed a presence beside my bed. I could smell flowers; the air was filled with the scent of lilies. Grandma's spirit lingered for a while and filled me with a sensation of peace and love.

"The next morning, I stepped in his room to make the bed. I found Grandma's pink nightgown lay under the covers on her side.

"Janet, remember your Aunt Emily in Ohio when she told us the picture of Grandma Marge on her bookcase lit up like an eternal flame? At first, she thought there was a fire in her living room."

"Oh, yes, Mom. And remember that one time we were taking dinner to Grandpa's house. You accused me of wearing your perfume. I wasn't wearing any. I know she was with us."

"I knew the fragrance was familiar. I was quick to judge you. Spontaneous moments can catch us off guard like that. My mother was looking after my dad from the other side. Ten years after she died, on the morning before Grandpa died, I drove to his home. I wanted to help the nurse change his bed. The television was off when I handed him his breakfast tray. He was bathed and had on a clean gown. He smiled and looked so content. Just before I left his room, I had noticed a reflection in the television screen. It looked like shadow wearing a hooded robe holding a sickle. I remember thinking *Not today. He looks so good.*

"It was that evening that his nurse telephoned to tell me he wasn't going to make it through the night. She told me to come at once.

"Grandpa passed away at home in his own bed. I was with him when he died. Since they have crossed over I know they are together. But thinking of last Mother's Day, I wonder if Grandma had anything to do with the spirit that has been with me."

"Mom, you know I feel Grandma's spirit when she is with me. I wish I could help you solve the puzzle. Right now I don't have any answers for you."

Janet looked at her watch, stood up, and placed her dish in the kitchen sink. "I have to go. Thank you for lunch, Mom. It's been nice reminiscing about past times and family. If you need to talk again, just call me. I love you."

Janet grabbed her purse and car keys. As we shared a hug, her long blond hair brushed the backs of my hands. "You know, Mom, October is only a few months away -- we could give ghost tours in this house."

"Janet, you're always looking on the bright side."

I gathered the remaining dishes on the table and put them in the sink, still thinking about Danny. The fall I was in New York,

LOVE FROM THE OTHER SIDE

Mother told me Danny had mowed our lawn and raked leaves through the summer. I received a few letters from him, asking me to come back. But I wasn't ready, and I needed more time. When I finally left New York to return home, I heard the news. Danny was married. Suddenly I had a feeling of loneliness; there was no one to share my life with. My thought was, *I guess I'll have to find someone else.* My stomach did a flip-flop, and my heart was heavy. I resigned myself to go on with my life.

Three months after I returned home I met Alex. After a few dates I felt so protected when his arms surrounded me. While I was growing up my family didn't have a lot of money. For our entertainment we spent weekends playing canasta, five-hundred bid, or nickel dime poker. When I met Alex, he was still in the Army and also didn't have a lot of money. He enjoyed playing cards with my family. They liked him and although I hadn't known him for long, it seemed like he had been with us all of my life. I trusted the light in his blue eyes. When he kissed me, there was this tingling on my lips and in my heart. The excitement stirred and aroused my senses. Now Alex was sick and tired all the time. It was only on a rare occasion I would find a screw on our nightstand.

My spirit guides must have known Danny's plan. It was the feeling of pure love while I was in the light, in my dream, and my faith in God that had kept me balanced and stable. What exactly was the lesson for this experience? If I knew, I could not remember. It felt like I needed to see this through to the end. There were still times when I felt lost and unsure of myself, but somewhere deep inside of me I still had that blind trust.

CHAPTER **13**

Automatic Writing

Later that evening, I reached for a legal pad on my nightstand. Sitting on my bed, I relaxed myself and picked up a pen. Alex was sound asleep beside me. Responses can come erratically and can be difficult to read, but things come through with patience.

"Hello?" I wrote. "Can we talk?"

"Yes we can talk, but you won't like what I have to say."

"Try me."

"You will be mine. Someday, you will love me. I don't want you to see Red."

"Is this about power and control?"

"Yes. I have the power, me."

"Some nerve. I guess there is nothing to talk about."

"So you want to talk to me?"

"You didn't finish your message."

"You will love me."

"What happened to your wife?"

"I don't love her."

"You don't spend thirty-six years with someone and not care for them."

"I want you."

"You repeat yourself a lot."
"You owe me a lot of years. I want all of them."
"Goodbye."
"Later."

SEPTEMBER

Alex awoke as usual at five a.m. He normally sat at the kitchen table laboring over a crossword puzzle, or reading his customer contracts. I usually slept until seven or seven-thirty. Alex knew when I was ready to take on a new day.

He must have heard me slip out of bed that morning and put my slippers on. He made a fresh pot of coffee and started to cook breakfast. The aroma of coffee brewing drifted in from the kitchen. I slipped into my favorite blue cotton robe. In the kitchen, Alex was standing by the coffee pot. Bacon and eggs were on the stove, and toast waited to fill the empty green and white plates on the table. I placed my arms around his neck and he put his arms around my waist. With my head on his shoulder, I felt his warm breath on my neck as we shared a hug. The bones in his shoulder blades…how thin he was getting!

"You feel so good in the mornings. Your hugs help me struggle through my day."

While pouring coffee in my cup I asked what his plans were.

"Well, we finished the driveway we were working on. I have a few appointments this morning. Not sure after that."

"Maybe we could shop at the mall. Could we buy a few things for the house?"

"It's a possibility." Alex kissed my cheek. "I love you, Carol."

"More," I said.

After breakfast, I dressed in brown slacks and a tan cotton blouse. This shopping time would be a good chance to talk to Alex. I knew we didn't have a lot of money to spend, but we really needed this day out to be together and just relax. For September, the day was unusually cool in southern Florida. Alex wore jeans, a yellow shirt, and his tan leather jacket.

"Alex, the cool weather feels good for a change after our hot summer. Do you think that leather jacket will be too warm?"

"The air conditioning is always so cold in the shopping mall. I don't want to get a chill."

"Really, Alex -- you must have Budweiser in your veins."

He returned home from his appointments around noon. Princess lay on the floor by the table. We ate sandwiches, fed Princess, and left for the day.

Walking through the mall, I broached the subject. "Have you answered any hang-up calls lately?"

"Yes. There was one this morning."

"You know, last May I told you I had seen my old flame at the Swap Shop. You've heard some of Red's messages to me. I want you to think about what's been happening around our house."

"You know I don't believe in that nonsense. You and your psychic friends -- I go along with it, but I don't believe it."

I folded my arms to keep from shaking and wanted to let out a scream. Why was he being so obstinate?

"Listen, when the hang-up calls first started, I was getting zapped. Now he just hangs up. You've been drinking a lot more lately. I can feel the negative energy. Your buttons are being pushed. You're not eating, and I know you're not feeling well."

"Nonsense," Alex retorted. "No one controls me."

I motioned toward the bookstore to calm down my emotions and anger, we entered. "Will you please pay attention to what has happened to us?"

I loved Alex; we shared the miracle of our two children and thirty-six years together. Why did he shut me out when I wanted to share my spiritual experience with him? Alex wasn't hurting only himself. He was also hurting me. In the bookstore, I wandered through the New Age section. I was looking for a psychic dictionary. Not finding one, I inquired at the sales counter. An elderly saleslady wearing a black dress checked the computer.

"Sorry, the only thing here is the Satanic Bible. Would you be interested in that?"

"No, thank you. That's not what I'm looking for." I felt a tinge of anger. Imagine a bookstore with a Satanic Bible and no psychic dictionary. What was the world coming to? It is my personal feeling and experience that spirits need healing as much as people do. I'd ask someone at The Haven for Spiritual Travelers.

I collected Alex from the Stephen King section, and we continued our walk through the mall.

"We could use a new teapot and some bath towels," I said. "You know, during Emily's last phone call she told me that my old flame, Danny, is dead. I need to go back north and find out what happened. He looked so alive at the flea market."

Alex held my hand.

"I get thoughts from Danny's spirit." I took a deep breath. "His energy is in the house." I looked down at the brown and tan tile on the floors that reflected in the store windows.

"I know about the calls. You've mentioned them to me before. I realize the stress you've had and I can see it in your expression. I don't know why I drink so much. I'll cut down on my drinking. You know I love you. As long as I'm here, I'll take care of you, Carol." Alex gave me a kiss on the cheek.

"I appreciate your chivalry, dear."

What can you do about an earthbound spirit? I can't just wish Danny away, especially if he is determined to stay.

I looked into Alex's blue eyes. They weren't as clear as they used to be. Something wasn't right.

"I want you to get a medical checkup."

"Don't be silly. All I have to do is quit smoking and I'll be fine. I'm not going to any doctor."

"I'll believe that when I see it." We had talked endlessly about Alex's health. He ignored my conversations.

That evening, Alex drank more than his usual amount of beer

until he was drunk and relaxed; then he passed out in bed.

I listened as Alex talked in his sleep. *"Do you want lovin' and freedom? Or fightin' and dyin'?"*

I knew Danny was talking through him. I wondered what he meant. *Lovin' and freedom, or fightin' and dyin.'*

Was Danny actually helping me through this crisis, or was he trying to hurt me? We weren't receiving any more hang-up calls. Now the phone would ring twice, and then quit before I could answer it. I knew it was Danny.

On September 29, I had a dream and was shown a key on top of a church altar. Then I heard a message. *"The key is love."*

I had thought about exorcism. I remembered the movie The Exorcist. The method seemed violent to me and without compassion. I knew his spirit was inside of me and I remembered the Danny I had loved. The message echoed in my mind. *"The key is love."*

I've always been told when a spirit dies it is supposed to go into the light. I realized that Danny needed help into the light. I wanted to help him find it, but it seemed that Danny did not want to go. He wanted to own, love, and keep me. His spirit was reaching out to me. My instinct was to do the right thing. I realized that his spirit came to me for help and I wanted to find closure, healing, and peace for him and me. I had to have been chosen to lead Danny to the light because of the energy we experienced and held. James Van Praagh titled it Unfinished Business.

CHAPTER 14

Karma

On Sunday evening, October 8, I drove to The Haven for Spiritual Travelers alone. Red began the service with his usual songs and meditation. He loved to tell his stories.

"You know there is a Spiritual Friend named Sandy who was having severe migraine headaches. It was ten years ago. I was invited to Sandy's home. Over a cup of tea we talked at her kitchen table. Then I told her it was time for me to leave.

" 'Well,' Sandy asked, 'when are you going to heal my headaches? All you did was come over sit at my table and drink my tea.'"

"Two days later, Sandy telephoned me. 'My headaches are gone,' she said.

"Just yesterday she called again and told me she hasn't had a headache in ten years. She wrote a testimonial about my healing. You'll find the letter is on a table by my front door. Do any of you spiritual seekers have any questions?"

No one raised their hand. Red looked over guests in his Florida room. "Well, I guess there's nothing on your mind. Why are you here?"

A well-dressed lady who looked to be in her forties raised her hand. "I'm here because my friends asked me to bring them. I don't believe in psychics or this kind of service."

I looked at the lady and thought she was so tight if she bent over she would crack in two.

Red reached toward the shelf beside him and picked up his Bible. It was well- worn, with Post-its on every few pages. He looked at the lady. "I don't care for the word psychic. We are sensitive. We touch, we feel, and we hear. This is all natural." He read from the Bible, Corinthians chapter twelve. "Now concerning spiritual gifts, brethren, I would not have you ignorant. Ye know that we were Gentiles, carried away unto these dumb idols, even as ye were led. ----- 4 now there are diversities of gifts, but the same spirit. 5 And there are differences of administrations but the same Lord." Red went on reading the Bible chapter. Then he continued talking to the lady. "Why are there so many churches and so many doors? If you don't come into this door you will not be saved. And if you don't come into that door you are a heathen. If you don't belong to so and so church I can't be friends with you. Churches and religion ruin people's lives with all of their rules and restrictions. I ask you why there are so many doors. There is only one God and one Book. People are ignorant of what is natural. People are afraid to touch each other. To hug a neighbor or hold the hand of a sick person, that is a natural healing. We are all born with a natural healing ability. It is to touch and to love. I have been asked to join healers on TV and on stages. I refused and I will not join them. They are about ego and greed. I like simplicity, to work one on one with a person."

The crowed gave a nervous laugh. "I bet when you leave tonight there will be many questions you wish you would have asked. You know my dear mother was a Cherokee Indian. She told me I have the gift of a Shaman, and if I don't use it in this life I'll have trouble in my next life. Well, I have enough trouble in this life. I don't need trouble in my next life. There are no stuffed shirts in my home. We are friendly and family. Let's prepare to meditate."

KARMA

The lady said little and kept quiet the rest of the evening. A short time later the crowd disbanded for a ten-minute break. We returned to the Florida room for billet readings. Red stood in front of his podium with a billet in his hand. His arms were folded in front of him, and a slight smile crossed his face. He scanned an envelope with his hand, looked up, and asked for "Jim."

"Hello, Red." A guy with dark hair in his middle twenties raised his hand.

"Let me come into this vibration."

"Yes, Red."

"You have asked about a person that has recently passed over. For this we have to go to the gate keeper. Because if they passed from a painful sickness they will be in a deep sleep and we will not disturb them. Do you understand?"

"Yes, Red."

"This person died of cancer."

"Yes, Red."

"They tell me she is at peace and will be happy."

"Thank you, Red. I feel better hearing that information."

My question was about Karma. I had been going to Red's meetings steadily for several months. But, I still had a lot of questions about my earthbound spirit, Danny, and where all of my experiences were leading me. Red held my billet in his hand and walked toward me. He channeled a deep-sounding voice.

"You will shoulder your brother's burden, dear. You will go where angels fear to tread."

Would that be Hell, I wondered, *with sickness and death?*

I felt tears well up in my eyes. I must have looked pretty sad because Red sent me a calming wave of energy.

He smiled and said, "Don't look sad. You will go by choice."

This message did seem to confirm my prophetic dream when I was lifted into the light last April. The spirit dream had said, "He has the gift. You will be all right. He loves you."

What was my other choice? I wondered. *To jump off a cliff?*

I asked him, "If I have to go where angels fear to tread, what will be my Karmic lesson?"

"You will learn faith instead of fear." He paused a moment and then looked at me, "You have a reason for being there. Polarity!"

I didn't want to believe I chose this, but I prayed God would walk me through it. I knew about positive and negative energy. For protection I wanted the Archangel, Prince of light, St. Michael there, too. He and his legion of angels defeated Satan and drove him out of heaven. I looked at Red and asked one more question. "What will it be like?"

"Like a child walking through a cemetery at night. There will be light on the other side." He smiled.

I admired Red as a friend and a teacher. He knew so much about my energy field.

"Isn't there another way for me to find closure to this? Can someone show me where to turn the corner?"

"Betty, no one can make your footprints for you. The weight of your feet imprints your footsteps. Follow your own path."

"Thank you, Red, but I'm Carol. Whoever Betty is, I hope she is helping me." He was always pulling names out of the air.

"We will figure it out, Carol. It looks like you folks are ready for a break. See you in ten minutes.

Spiritual Friends adjourned and retreated to the sitting room for coffee, cake, and tea. A plate of brownies sat on the table beside a pot-bellied stove.

"Red, this round table looks really old, is there a story attached to it?" I noticed the legs under the tablecloth.

"My brother gave me this table," Red replied, "It belonged to slaves from an old house in Georgia. There are burn marks on it from fire."

"I bet there is a good ghost story that came with that table."

"Yes, there is. Some day I'll tell you about it. Here, have a brownie -- I made them fresh this morning." He handed me a small paper plate with a brownie on it.

I asked Dawn, Red's office assistant, if she knew where I could buy a Psychic Dictionary. Dawn told me she had a copy of the dictionary.

"I'd be happy to lend it to you."

Just then a lady standing in the parlor doorway let out a scream.

Red hurried over to her. "Oh my God. Are you all right?" He put his hand on her shoulder.

"I'm so sorry." She looked embarrassed. "An icicle just passed through me."

We all laughed and I said, "It's nothing to worry about, dear. It was just a spirit passing, wanting to get through the doorway."

Dawn stepped into the office and retrieved a thick blue book. When she put it in my hands, I couldn't believe a book like this had even been published! It was the size of a telephone book. I looked at the title. **The Donning International, Encyclopedic Psychic Dictionary**. The author was June G. Bletzer Ph.D.

Under my breath, I said, *Thank you, God*. Immediately, I sat down and looked up the word Karma.

> *"This principle is governed by one's entire activities, thoughts, and emotions throughout all incarnations. For every action, there is a reaction. Karma represents the sum total of the causes one has set into motion in past lives, making the pattern for this life and for future lives. This pattern can be changed, and rearranged purposely through preplanned deeds, acts, and attitudinal changes."*

I could have sat in The Haven all night and fulfilled my questions with the knowledge I needed.

Then I looked up, POLARITY and found out it is a balance of opposites.

> *"The positive or negative state in which a system or an entity reacts to a magnetic or electrical field, and constantly seeks to balance 2.(esoteric) everything has a positive and a negative aspect which is necessary for balance. This balancing of opposites works simultaneously with the vibrations and human thought."*

I retrieved my purse from a corner in the sitting room. Red stepped in front of me and placed his hands on my shoulders. He looked into my eyes and smiled. I felt friendship, tenderness, and compassion from the light in his eyes. We shared a hug. I reverently held the dictionary, stepped through the doorway, and walked toward my car. While driving home that night, I was stopped at a red light on Davie Boulevard, waiting to turn left onto I-95. My truck started to rock back and forth.

"Who is rocking my truck?" I said in a loud voice. *Could this be jealousy from the spirit?*

My truck immediately stopped rocking. I knew a personal struggle was coming. And if Red's message was true, it was not going to be easy. Independence, power, and freedom of spirit would be part of my fight. Remembering my dream, I wasn't sure what I was supposed to understand. *"He has the gift. He loves you."*

I was fighting to keep my balance and sanity. I needed to keep taking my evening walks. Danny's spirit knew Red could release him to the light. When we were growing up, meditation circles didn't exist. Now Red had led me to a greater understanding of spiritual awareness.

Alex and I felt so tired most of the time. I knew Danny was draining Alex's energy. I couldn't sleep at night. I was not working full time, and still I was so tired. Dawn, Red's office assistant, told me I could stop the tiredness, the spirit visits. Red told me I could end it at any time. I wanted to. I thought of a regression. People I knew had closure that way. I also knew it was expensive. I would not ask if I couldn't pay. I was not the only one involved here. I didn't believe I asked for this. It was thrown at me. When you turn a corner and a door slams you in the face, what are you supposed to do? I wondered if Danny's family talked about spiritual happenings. My family had a whole history of strange stories.

Some say I created this, but they can't tell me how to solve it. My spirit guides told me to follow my path. To everyone out there who does not believe in spirits, just take my place. You may then experience my abundant mystery and confusion.

CHAPTER **15**

The Accident

I had started going to Mass on Sundays, after my experience at the Swap Shop flea market last spring. My childhood roots were planted in my Catholic faith. I received an inner peace by attending Mass and Holy Communion. Although it was the same church I attended when we first moved to southern Florida, I was not familiar with the congregation because I missed close to fifteen years. I had thought about talking to a priest, but they were so busy with the schoolchildren and parents. I had made two appointments on separate occasions. Each time, the priests were double-booked or had an emergency. I really didn't think they would help me with my earthbound spirit. I knew the church did exorcisms. I felt I was given my message for a reason when the spirit lifted me to the light last April. I wanted and needed answers.

There are so many wonderful positive things that can be done with psychic energy, Healing, love, the God self. Miracles and wonders of spiritual energy abound. Why do people have to use it in negative ways? I had to admit I was romanced by the mystery of it all. I was also scared. Like the warrior who goes into battle, I had no idea how I would come out of this. But the situation was thrown at me and it wasn't going away.

Still I asked, "Dear God, why me?"

On the twelfth of October, I awoke from a deep sleep. I had had a dream about my son Jerry, something about his blue Ford truck and a bridge, like a long blue fish sailing under darkly moving water. I'm grateful for my dreams and premonitions, although I don't always understand the symbols my spirit guides show me. Most of the time, I cannot change the outcome of a situation. I believe everything is planned. The lessons teach, and help us become better-balanced beings of light. I wasn't sure what might happen with Jerry and his truck.

Dear God, I was told you wouldn't give me anything I couldn't handle.

I telephoned Jerry at his apartment. "Be careful! I had a dream about you and your truck."

"Mom, stop worrying about me. I'm a big boy now. I'll be careful."

I just knew whenever Jerry was with his friends, all my warnings went out the window. Although he was thirty-four years old he acted like a teenager. On Thursday, Halloween night, the phone rang. I looked at my bedside clock. One a.m. We were a one-phone home. I hurried to the kitchen and reached for the telephone. I held the receiver. My heart was in my throat. My stomach was in knots.

"Hello?"

"Hi Mom, it's Jerry. I'm at Imperial Point Hospital. I've had an accident."

"I'll be right there."

Tears welled up in my eyes. As I hurried and dressed, I kept thinking, hoping he was all right. Jerry was well enough to telephone me. There was no use in waking Alex. He had an early morning appointment, and had passed out earlier. He wouldn't be any help. I was so scared as I parked our work truck in the hospital parking lot. A Broward County Sheriff -- Officer Dolan -- met me at the door.

THE ACCIDENT

"Mrs. Shimp?"

"Yes," I answered. He reached for my hand.

"Your son will be all right. All he needs are some stitches to the side of his jaw. He was drunk. He tried to take out a cement wall. A wall was built recently to block off the decaying bridge."

"Jerry drives that road all the time -- why didn't he see the barricade?"

"There were no reflectors by the bridge. His truck was totaled. Your son was very lucky. Anyone else might not have lived through an accident like that. Being drunk and relaxed might be what saved him."

Officer Dolan looked to be in his fifties. His hair was the color of salt and pepper. He was gentle and seemed concerned. I thanked him for his sensitivity. Then, an emergency room greeter ushered me into the room, where I found my son in a hospital bed. A nurse informed me the doctor wanted to keep Jerry overnight for observation. I stood beside Jerry's bed, smoothing the blond hair off his forehead.

"You are so lucky," I told him. "Grandma and the angels must have been with you."

"I'm sorry, Mom. I just got carried away." He tried to smile. "I didn't know they built a fuckin' cement wall in front of the bridge."

I figured Jerry was more upset about wrecking his truck than he was about being drunk. I was grateful he was okay and stayed with him for a while, held his hand, and told him about the police officer's concern. I told him I would bring clean clothes in the morning, when I came to drive him home. I arrived home around four a.m., and threw my truck keys on the table. All I could do was sit on a chair, and hold my head in my hands. I was so tired and depressed. I knew Jerry wouldn't change. He didn't listen to me. When he was with his friends, it was one big party.

November

The following morning, November first, I stepped into the kitchen. Alex was going over old tile contracts. We hadn't had any new ones lately. I poured my coffee and sat on a chair at the table across from him. I felt despondent as I looked at the dark circles under his eyes, and at his thinning hair.

"I have something to tell you, Alex."

"What happened last night? I heard the telephone ring."

"I was at the hospital with Jerry. He totaled his truck."

"Hospital? What the hell happened?"

"He drove into a cement wall that was built in front of the bridge, just off the Brickell Avenue extension. I'm glad the accident wasn't far from home. The doctor said he will be okay. He needed several stitches in his jaw. The doctor wanted to keep him overnight for observation."

"Well, thank God he is okay. I do worry about the children, you know. Why didn't you wake me?" Alex looked dejected.

"You were passed out and I knew you had an early appointment this morning. You might worry about them, but you're never around when you're needed."

"That's not fair, Carol. I work hard to keep this family going. You don't know what I have to put up with out there on the road."

"Well, I didn't bother to wake you last night. I went to the hospital by myself. You seem to have passed your drinking habit on to Jerry. I'm probably going to lose both of you. Oh, and by the way, you were never at the hospital with me when Janet was six years old and was diagnosed with diabetes. Then there was the time she had an insulin reaction in grade school..." I said. I knew I was putting it on.

"Carol, I was on a job site. How was I supposed to know?"

"I called and left messages. I guess you didn't get them. Alex, I'm sorry to vent my wrath on you, but I'm tired of dealing with emergencies alone: the bank forecloses on the house, no work, and no money. It seems like all you do is drink beer."

"The beer helps, Carol."

"I'm going in the bedroom to meditate. It seems my spirit guides are the only sense I can make out of my life these days. I love you, Alex. But what has happened to us?"

I finished my coffee and left for the bedroom.

My eyes welled with tears as I faced our window toward the east. After I said my rosary, I dressed, picked up the truck keys, and drove to Jerry's apartment. I had told Alex I wouldn't be long. He needed the truck for work. While gathering clothes in Jerry's apartment, I opened the refrigerator to check for food. Of course, it was almost empty. On the way home from the hospital, I told Jerry we would stop at the grocery store. I had made a list.

"I don't need anything, Mom; I've been through this before."

"Well, you are my son. I love you and I worry about you. We will pick up soda, lunch meat, bread, orange juice, and a few frozen dinners."

"Thanks for picking me up, Mom. I called my boss from the hospital. He told me he knows someone in the roofing crew who will drive me to work. Mom, I'm going to fix my truck. You'll see. It'll be just like new."

"Well, it needs a new front end and I don't know what else. I know you've always done your own mechanical labor. I guess you could find parts in the junkyard. Jerry, I hope you have learned a lesson from this. I have no money to help you. Jerry, you are going to have to be responsible for your own actions. Right now, I don't even have the energy to be angry with you. Between Dad's health and the bank notes, I'm exhausted. I just want to tell you: don't drink and drive!"

"Sure, Mom. Not anymore. I'll find a ride to court and DUI School. You know it was dark out. I was coming home from Ray's Halloween party. There were no lights or warnings by that bridge. It wasn't there two days ago when I drove down that road."

I pulled up to Jerry's apartment and helped him carry groceries inside. I made it home in time for Alex to keep his appointment.

When Henry and Jerry's friends heard about his accident, they made light of it. Do guys ever take anything seriously?

I heard one say, "Hey, Jerry, I hear you're driving through cement walls."

I had wanted to thank Officer Dolan personally. We later found out it had been Officer Dolan's last night on the job. He had retired.

CHAPTER **16**

Where Angels Fear to Tread

It was November fifth, my mother's birthday. Her spirit came to me often. I could feel her energy play with my hair. Sometimes, it felt like a silk scarf across the back of my leg. She always came to me with love energy and help from the other side. Ninety percent of my automatic handwriting was done at bedtime when I was relaxed. Alex was asleep beside me. I held a legal pad and started to write. The energy was mixed.

"Hi Mom, Happy Birthday."

"*Yes, thank you. Love you too.*"

"Mom, it's so great to have you here. Were you at Red's with me on Sunday?"

"*Yes. Red loves you and will help you. But you will want to see your Danny.*"

"Mom, I remember in my dream last April, I was told Danny had the gift. But I'm wondering if my subconscious was not programmed by the wrong energy? And why do I have to go where angels fear to tread?"

"*You have to. It is the plan. You will be okay, because Danny loves you.*" I felt a change in the energy. It was subtle, like a light switch turned on in a dark room.

"Now what?" I asked. My hands were warm. I knew he wanted to talk. Then his words appeared.

"You will not go to Red's anymore." That was Danny.

"Why me?"

"Because I love you."

"You're trouble!"

"I am not trouble, because many people will want my help. Do you want to see me?"

"Yes. I want to kick your ass for screwing up my life!"

"I will let you do your thing in the future."

"Were you with me in the grocery store last night?"

"Yes. You looked so good I wanted to screw you."

"It's time to say good night. I need rest."

"You will love me."

The following evening, November sixth, I closed the front door behind me. I liked to walk in the evenings after dinner and dishes were finished. The air was cooler, and I enjoyed the beautiful sunsets. Nature was multi-tasking with my senses. I was blessed with aromas of fresh cut grass, jasmine, and late dinners being prepared. On my walk that evening, I felt Red's energy. He was with me in an out-of-body, astral projection. He often came to me that way. He sent me a thought. *You know I am your people and I will help you.*

I responded in thought, *Thank you, Red. You're welcome anytime.*

I turned the corner and started to walk down the path toward home. Just then, a blue Chevy drove slowly by me. I glanced at the driver. It was Danny. He looked at me and smiled. How, I wondered, could his spirit walk into an individual and materialize? I just kept walking,

He had some nerve.

Alex was going downhill fast. He was having problems with his bowels. At times, he could hardly make it to the bathroom. We continued to receive foreclosure notices from the bank. I was scared, not knowing where we were going to live or what our future held. If we lost our home, I thought I could deal with it. It was small and needed a lot of repair. I thought, dear God, if there are any miracles left in heaven, I could sure use one. I just prayed we would get through the holidays and then tackle the New Year. We did finally apply for Medicaid. I had put several job applications in the local grocery and discount stores for employment.

I decided to write my friend, Mark Dodich. We didn't write often. He was living on the West Coast, in Oregon. I knew he was busy. I needed more answers and some kind of confirmation. I trusted Mark, and he gave me comfort.

10 November 1996
Hello Mark,

I know I haven't been keeping in touch. My life has been hectic. I'm hopeful this year has been better for you.

I have been really challenged recently. When I wrote to you last June, I had asked about my hang-up calls and being zapped over the phone. Well, my earthbound spirit is still with me. I'm looking for answers. I'm optimistic in finding closure to my past and future.

Time and energy flowed easily for me when you lived in Florida. We had good times. But I know you have a better following in Oregon.

I need a spiritual and emotional lift up. Could you please help me, with a reading of my planets for the rest of this year and maybe next? I have faith and trust in your predictions.

Hopefully we will be able to stay here until after the holidays. The bank is foreclosing on our home. Alex has not been well for several years now. I'm looking for work. I applied at the local grocery stores, also Sears, Target, and a few other places. Janet wants to be single again. She is not happy in her marriage. Jerry is trying to repair his truck after a bad accident.

I'm still writing in my journal. Some day or year, I would like to edit it for publication. I believe my story will have a message for readers. I'm hoping it will provide insight to similar problems. I'm enclosing a small check. I wish I could send more;, fifty dollars isn't much for all of the work and research you do. Thank you for all of your insight, love, and light. I wish you well.

Blessings,
Carol

I read the letter over again before mailing. *We will be moving,* I thought. Could I deal with losing our home?
I received an answer from Mark the following week.

Dear Carol and Family,

Thank you for your check; it will come in handy for Christmas. Well, it sounds like you have had a busy summer, or maybe a busy year. If you believe all of the prophecy and Psalm 97, it is just getting started! Almost everyone I know is experiencing challenges of one form or another. Those who stay flexible and trust seem to be doing better than others.

A big thing coming up in 1997 has to do with your north node. This is a mathematical point relating to the moon. Eastern philosophers call this Dharma. Dharma is the path you're going into in this life and relates to your soul path.

Currently, the north node is traveling through your fourth house.

This means you are in the process of reviewing your security issues. This can be something tangible like the home you live in, or it can relate to maternal mother (or sometimes paternal father) issues from early childhood. Now is a great time to look at what you learned the first seven years of your life, regarding what made you feel secure. With Virgo ruling your home environment, a certain amount of order and cleanliness to the home would be one issue. No doubt, this also could bring insight to your writing, in the home.

By mid-1997, intuition and spirituality in the home could really come to the front of your life. Positively, it helps you get in touch with your emotions, spirituality, and provides insight into soul purpose. Negatively, it can send you off in wide mood swings. No matter what, it requires you to look at your inner security needs.

All in all, it looks like you are moving in the direction of some very positive changes. Since several of the planets move slowly, do not try to push the river. The Universe has a divine plan. We must tune into our inner knowing to know what is right at what time. That Mars and Pluto energy you are experiencing probably wants to force the pace. It's not a good idea.

Blessings,
Mark

I had to pull out the blue book, the ENCYCLOPEDIC PSYCHIC DICTIONARY, to look up the definition of DHARMA.

"Yoga dhri, To hold, that which upholds. An unknown quantity or principle by which the universe is upheld; that which pervades the whole universe and regulates its harmonious action; that which is the cause of the

universe, its preservation, and its final dissolution or absorption into the Supreme Scheme of things. 'MONAD, SPIRIT.'"

I was still confused. I wished I could say I was a scholar on this subject. But I felt so inadequate. I realized more about Danny's struggle, and my struggle to understand. But I could see in Mark's letter that I was moving in the right direction.

CHAPTER **17**

The Dream/Thanksgiving

There were nights when I slept beside my husband and felt the presence of Danny's spirit lying with me in our bed. His spirit didn't upset me. He was comforting, like a sister or brother just wanting to share love energy. He didn't do anything. It just felt like he wanted to be there. I received a thought message.

Danny showed me a tombstone. Then he held a sledgehammer. He was hitting the tombstone with a hammer. Then we were in 1959, our hometown on the street where he and I both lived. I had a vision of the car he wrecked. He handed me a red rose.

I returned a thought.

You are asking for my forgiveness, but you suffered the consequences for your actions. I have always loved you, but it was that damn temper of yours. Yes, I had cold feet and left town. You decided not to wait for me, but I have a family now, a full life. I cannot give it up and return to the past.

I felt a light pressure from his presence. It was a hug. I smiled. I felt him hesitate. He knew I had smiled. *Yes, you made me smile.*

After a minute, Danny's spirit left.

It was close to Thanksgiving 1996. My family watched the Miami Dolphins on TV. Janet had made orange and teal penalty flags to throw at the television when we disagreed with the officials. A large bowl of popcorn, and beer and soda cans sat on the coffee table. I was not into football, so bored beyond belief, I decided to visit the neighborhood shopping center and look for a new blouse. I entered a small clothing store and chose a few blouses off the rack. In the dressing room, I decided on a blue cotton blouse. I like cotton. It's always cool and comfortable. After I dressed in my original outfit, I heard my mother's voice.

"Comb your hair."

I was surprised she was with me. I had become so used to her spiritual energy that at times, I didn't realize she was with me. My mother's spirit was very busy.

I looked in the mirror combed my hair and said out loud, "It's all right."

I paid the cashier at the desk. She was young and looked bored. I was the only customer in the store.

There was a neighborhood bar in this particular shopping center, frequented by many of the locals. While I was walking through the parking lot, a car had pulled right in front of me and parked. Danny left the car and headed for the bar. He seemed shocked to see me, but nodded slightly in my direction. I returned the nod and kept walking, because I was also shocked. Danny looked so alive! Was he alive or dead?

I wondered why my mother's spirit wanted me to comb my hair. Did she know I would be here and Danny would show up? Mother knew Danny in spirit. Her messages to me had been helpful.

The day before Thanksgiving, Janet came to the house to help bake our traditional pies: pumpkin, apple, and cherry. With our hair in ponytails, and dressed in old T-shirts and jeans, we started. We made everything from scratch, and did our usual laughing, bitching, and conversing. We discussed Alex's problems.

"What are my options Janet? Rehab, send him away, or just let him drink himself to death?"

THE DREAM/THANKSGIVING

"Mom, I'm just so worried about you. What are you going to do?"

I looked at Janet. "Really, I'm at my wits' end. Your dad is too young for Medicare. I wish we'd hear something from Medicaid. It's been months since we applied."

"Dad embarrasses me, Mom -- I get so irritated with him. Why does he drink?" She dropped a knife on the floor, retrieved it, and placed it in the sink.

"Janet, it's a sickness -- and If he doesn't want to stop drinking, all the rehabs in the world won't help. I can't just leave him. I love him. He told me if I put him away, he would commit suicide. I know Dad is having a problem with his lungs from smoking, and now he's got bowel problems, which I'm sure are related to his liver. I don't have any answers." I choked back a sob. Janet reached for my hand and squeezed it. We shared a hug.

"Mom, I'm here for you."

Baking and sharing the holidays kept our family close. We discovered a lot about each other.

I placed two apple pies in the oven, and started to peel more apples for applesauce. Janet opened the kitchen door to let the dog out. She shrugged her shoulders and looked at me with troubled eyes. She started to roll out pie crust for her favorite, cherry pie. I had hoped our traditional dinner and baking would release some of the stress we were all having.

"You know," I said, "the last time I saw Red, he said my past would catch up to my present and my present would be my future. I'm sure there is a plan somewhere. I have to keep my blind trust and faith."

"Wish I could help you, Mom, but I've been busy and had problems lately."

I looked at her as she continued, "Henry and I have a commitment at the Star Trek convention coming up at the Hilton. I have inventory at the shoe store. You know I have been there for fifteen years now? Then we have a meeting next week with our Klingon shipmates. There aren't enough hours in the day." She popped a cherry in her mouth. I looked at my daughter, and marveled at the grown, dependable, and caring woman she had become.

"I noticed some tension between you and Henry. Do you want to talk about it? "

"Mom! You know we have been married for fifteen years now." She hesitated, looked at me, and then told me, "But I have met someone at our Klingon meetings that I really have feelings for. In all the years I have been married, I have never felt like this about anyone. I'm not happy anymore." She looked upset.

"I know sixteen was a young age to get married, Janet, but you were so sure of yourself. Please don't do anything you will be sorry for. You and Henry have always gotten along so well together. This feeling you have for the other guy could pass. Give your marriage a chance to work out. I haven't been happy for a long time. I don't know where my future lies."

She paused and looked down at the floor for a moment. "I'm just so unhappy."

"Give it some time. Don't do anything rash."

The men came home from work. Jerry and Henry had helped Alex finish a job. All three of them came in the kitchen and sniffed the air appreciatively. Henry was always a tease, and I watched as Henry picked up a fork and carried a whole pie into the TV room and sat down. Henry's father was part Crow Indian and his mother was French. He identified with his Indian heritage: tan skin, black wavy hair, and blue eyes. His cute dimples always gave him a mischievous look. I hoped Janet would work things out.

"Bring that pie back in the kitchen!" my daughter yelled.

Henry came in the kitchen with a sad look on his face and placed the untouched pie on the table.

On Thanksgiving Day, we enjoyed our traditional dinner. I removed, from a drawer in the kitchen cabinet, a white linen tablecloth we had used on many occasions. Janet set the table.

"I always loved Grandma's china," Janet said, admiring the plates while she placed them on the table.

"Yes, it was Grandma's favorite. When she saw the blue and white Currier and Ives set, she mentioned to her sisters how much she liked it. The family pitched in and bought it for Grandma and Grandpa's twenty-fifth wedding anniversary. When Dad and I move, Janet, it's yours. Boxing it and carrying it from place to place is a struggle. I need to lighten my load."

"Oh, thank you, Mom -- I'll even have holiday dinners in my home. Just so I can show it off."

I said the blessing before we ate. "Bless us, oh Lord for these Thy gifts which we have received from Thy bounty, through Christ our Lord. Amen."

My husband was in his glory as he looked at our family that surrounded him. "All of this food and football too," Alex gloated. He started to carve the turkey while Janet passed mashed potatoes.

Jerry did his usual tormenting. He put pepper on the dog's nose to make her sneeze. I had wanted to ask his girlfriend for dinner, but they were having one of their disagreements. Their relationship was always on or off.

Alex said, "You should have seen Princess last night. Her feet were going a mile a minute, doing circles in her sleep. I'll bet she was chasing a rabbit. Make a good Duracell commercial, the battery that never runs down."

Just then, the telephone rang. I answered it. I let out a sigh of relief; it was not a hang-up call. It was Jerry's girlfriend. She told me her water pipe had broken and water was all over the front yard. I looked at Jerry and repeated her message.

"Good!" he said. "I'm not talking to her, and I hope she had PMS and a hangover when it happened!"

I smiled to myself, and after telling her to call the city, I hung up the phone. I believe in unconditional love. It is the overview of our experience with Spirit. I tried to guide my children morally and spiritually but they also learned from some of their mistakes. That is why we are here on this earth plane, to learn. If anyone could keep me grounded in reality, it was my children. I looked at Janet, her blond hair and blue eyes, and Henry with his wavy hair and dimples. I sensed sadness between them.

Dear God, I was told you wouldn't give me anything I couldn't handle.

After the dishes were done and the football games were over, the family left for the evening. Alex was asleep on the sofa with our Princess at his feet. I decided to leave them there.

It was close to 10 p.m. when I decided to walk around the neighboring block. There were times when it was raining, or I had been out for the evening and arrived home so late that I didn't take walks. Those were the times when I would exercise in the house. I would do situps, or arm and leg stretches. But tonight I wanted to feel the fresh cool evening air and get some stress relief after the long day. A few homes on the block already had Christmas lights glowing in the windows and in the yard. The quiet night relaxed me. The streets had little or no traffic. I walked toward a corner in the road and I saw a large black dog sitting in the middle of the street. The more I studied him, the more it looked like a Devil dog. His red eyes were lit -- they were not normal, and his stare was fixed on me. Mentally, I put my guard up. I wasn't afraid. I was irritated. Instinctively, I recognized this dog, just sitting there, represented a message from Danny. I couldn't believe this person I used to date and almost married was on the dark side. I kept walking, trying to ignore the dog as he just sat there and stared at me. I passed him and halfway down the block, I glanced back to see if he was still there. The dog was gone.

The Catholic Church had always taught me that if I was good and lived right, God would take care of me. Now I was seeing a dark force entity, and wondering what happened. I realized that even with my experience, I had limited knowledge and distorted perceptions of the spirit world. How could I be so dumb and leave myself so vulnerable?

THE DREAM/THANKSGIVING

I returned home, showered, and went to bed. I reached toward the nightstand for paper and pen. I was hoping to reach my mother.

"Hello?" I wrote.

"Yes, I love you." It was Danny.

"Love is a wonderful thing."

"Yes! You are in for a big surprise. As soon as the weather changes you will go to our home town and find out the truth. Then you will love me."

"I have always loved you."

I felt a change of energy, like a small electric shock. "Mom?"

"Yes, it is me."

"How strong are my powers to see through the veil and other worlds?"

"You are special. You have the power to see. Danny is on the other side and he needs your help. You will know the truth and it will be soon."

"But Mom, when will I know for sure? "

"You have a gift. Love is the key and the way is with us."

"Mom, we are in so much trouble. I love Alex, but he is not well and the mortgage company is foreclosing on our house. I'm so confused. I take one day at a time."

"I know, but he will be with me."

I was not shocked by my mother's message. "He will be with me," knowing how long Alex had been drinking and in ill health. What I didn't know was how or when. I know the infinite and eternal healing power of unconditional love.

"I pray I can sort all of these messages out."

"You will. We all love you in the spirit world and on the earth plane."

"Thank you for your help and messages. I love you. Good night."

The morning after Thanksgiving, I made our bed. As I brushed my hand over the sheets, on Alex's side of the bed, I discovered a tiny white moth. It was the size of a nickel. I had no idea how this tiny fragile creature had found its way into our bed. My hand rested on the sheet and the moth walked onto the back of it. I carried it to the front door, held my hand outside, and waited for it to fly away. At first, it didn't want to leave. I moved my fingers up and down, and then the moth flew away.

CHAPTER **18**

December and Christmas

In early December of 1996, I attended The Haven alone. On the last few Sundays, Alex was always watching football with friends. Red started his service with singing and meditation. He wore his usual jeans with a red sport shirt. The light reflected on the Kokopelli charm that hung on a chain around his neck. His dog was lying on the floor by his feet. Red reached down to pet Morgana, a five-year-old Golden Retriever. She always greeted and showed affection for everyone.

He looked at her large brown eyes.

"You know, when I come home with a truck full of groceries this dog thinks I'm a great hunter.

"Yesterday, a long black limousine pulled up in my front yard. The driver wanted to know how much I charge for a healing. I told him it was cheaper than a funeral." Red paused for a moment, then continued. "Three friends died in a car crash. They met Saint Peter in heaven at the gate and he asked all of them the same question. When you were lying in your casket, what was it you wanted to hear your friends say about you? The first guy said, 'I wanted to hear that I was a great husband and father.' The second guy said, 'I wanted to hear that I was a great teacher and made a difference in the children of tomorrow.' The third man said, 'I'd like to hear them say: Look, he's moving!'"

We had a meditation, and then our ten-minute break. Red started reading billets. He held my billet in his hand as he walked toward me with his arms folded. Red smiled and looked me in the eye. "I want to tell you your timing is getting better."

I was completely embarrassed.

Red looked at his spiritual friends. "And I bet you would like to know what we're talking about."

"Thanks, Red," I answered.

I was embarrassed that I was embarrassed. When I meditate, it is my private time with God. I knew Red was with me, or passing by in the etheric field. That was fine with me. There were a lot of entities around, including Danny. When you deal with spiritual fields, out-of-body experiences and so forth, there are no secrets. As long as they were helpful, I didn't mind. How could I go through this incarnation and not feel the energy around me? I know that our planet and the universe have many levels of consciousness, spirit, and people, aliens, dreams, thoughts, and God. The energy is amazing. If I had to live in a cocoon on one level, it would leave me hollow inside. Just the thought is depressing.

But when Red mentioned my timing in front of everyone, he took me by surprise.

Red told everyone he meditates at six every morning. Spiritual travelers are welcome to join him in their special places at that time. I had been meditating around eight a.m. But for the last few weeks, I couldn't sleep. So my time had moved up to seven a.m.

How did he know? But I knew that he knew.

Red went on to answer my question.

I'm still looking for answers.

"I see a can of food. There is nourishment for you, but the lid is open and jagged. You can be hurt." Red looked upset. "You have to go through the lepers to get to the lilies. Even a snake can be charming, and you know who I'm talking about."

I was aware of Danny's presence. I could feel his spirit standing beside Red in the etheric field. I watched Red tilt his head to the right.

Red was talking to Danny. "My first wife left because of family interference. I let her go."

To some, it might seem like he was talking to thin air. But I knew Red was trying to reach a truce with Danny. He must have made an impression. For two weeks, the energy was normal around our house. No one could have helped me the way Red had.

Two weeks after I saw Red, I awoke from a deep sleep. I had received a mental message in my dream. *IF YOU THINK YOU KNOW THE END OF THIS STORY, YOU ARE SADLY MISTAKEN. BLANK WHITE PAGES!* Here we go again. I knew Danny's spirit was not happy with Red.

I felt that my messages from Spirit in automatic handwriting had been candid and unvarnished. I would go back days later, re-read them, then pick up the messages, try to figure out the puzzle, and then tune into the energy that came with them.

Two weeks into December 1996, Alex was napping in the late afternoon. I opened the bedroom door and stepped in to wake him for dinner. He lay on the bed in his cargo shorts and blue T-shirt, half-asleep. *Dear God*, I thought, as I noticed how loose his clothes were and how much weight he had lost. I sat on the bed beside him and gently ran my fingers through the stubble on his chin.

"How did you get that scar under your eye?" I asked him.

"I don't know. I have scars," he said.

I softly touched the crow's feet under his eyes. "Scars and wrinkles give you character," I told him.

"I have character." Alex smiled.

I gave him a kiss. "Thank you for the red roses you brought me last night. It was a nice surprise. I put them in a vase on the counter by the kitchen window."

"I love you."

"More," I said. "Hot dogs are on the stove and baked beans are ready. Come eat dinner with me."

"You talked me into it."

One day Alex said to me, "I'm so tired. I just don't want the problems with life anymore. I want you to look for someone else."

I put my arms around him. "Don't talk like that. I love you."

"More," Alex said.

He brushed the hair from my shoulder and looked into my eyes. "Carol, I'm not getting any better. I'm so tired all of the time. I have disappointed you and I feel so crummy. I want you to be happy."

Choking back the tears, I held him tight in my arms. "Alex, I've cried a lot this last year. Before long, we will lose the house; Janet's talking about a divorce, the accident with Jerry's truck. But you are my reason for pulling through these crises. I couldn't do it without you."

"Carol, you have a strong inner spirit. I'm holding you back."

With the back of his hand, he wiped a tear from my cheek. We held each other for a while. I did not want to give up Alex. He was the whole of me. I shared his breath, his warmth -- and for many years, our happiness.

In my mind and heart, I knew he was suffering. Tears welled in my eyes. *If you want to free your spirit, I won't hold you back. In my heart, I know you will always be with me.*

One week before Christmas, we pulled multicolored lights out of a box and hung them in front of the house. I had a few dollars for cookie supplies. With Janet's help, I was going to bake our traditional Christmas cut-out cookies and nut rolls. We had always boxed them as gifts for friends and neighbors. I hoped our positive energy would bring some peace to everyone.

Janet and I stood at the supermarket's checkout counter with ten pounds of sugar, flour, butter, eggs, sprinkles and colored sugar, decorations for cookies, yeast, nuts, and other items on our list. I noticed a man standing near the produce section. He was

DECEMBER AND CHRISTMAS

slim and looked to be in his late forties. He had dark hair, and was wearing green work pants and a sport shirt. He stood there with his arms folded, staring at us with an arrogant smile on his face. He nodded toward me. I thought about the statement Danny had made in my automatic handwriting.

You owe me a lot of years and I want all of them.

I had to ask myself, did his spirit walk into this person?

So he wants Christmas cookies, does he?

The following day, Janet and I were in the kitchen dressed in our usual baggy white T-shirts and old blue jeans. The temperature was in the high fifties, cool weather, and our windows were open. Our hair was tied back in ponytails. I was mixing ingredients in a large bowl. Janet was greasing cookie pans and wiping the cookie cutters clean. While working she sang,

"On the first day of Christmas, my true love gave to me...a hardon in his blue jeans."

Jerry walked in the front door and overheard Janet singing. He teased, "Mom! And you holler about me! She's just as nasty as I am. Twisted sister!"

"Well, you didn't get it from my side of the family."

I looked at Jerry's ragged T-shirt and shorts, and his short blond hair with tangled curls.

Guys always seem to be blessed with curly hair. A fresh, cool breeze from the open window grazed my face and hair. One of the great wonders of living in southern Florida is being able to open windows in the wintertime.

"The smell of our baking will bring all the neighbors to our front door." I smiled to myself.

Jerry put his finger in the cookie batter, swiped a fingerful, and ate it.

"Hey! Stop stealing the cookie dough!" Janet yelled.

He took a glob and put it on her nose. She wiped her hands on his shirt. He put his hand in the bag of flour. Before Janet turned around, he threw a handful over her.

"Stop it!" I yelled. "I'm not cleaning this mess up. I can't believe you two are in your thirties. Just look at the flour on the floor and the counter."

Janet took a handful of flour and threw it at me. The three of us stood there and looked at each other for a moment. The taste of flour in my mouth was pasty. Janet looked like a clay statue with blue eyes. Then we all burst out laughing.

"You two have never grown up. I remember a day when we were at Grandma's house. Janet, you asked for a soda. Grandma had only one can in the refrigerator, and she said you had to share it. So Jerry took the can and poured half of it over your head." I looked at Janet.

"Yes, and I remember when Grandma in her blue flowered housedress was chasing Jerry around the backyard with a broom." Janet looked at Jerry and smiled mischievously.

"Mom, look what she started. I don't have to listen to this."

"You children are the reason I'm never going to grow old. Janet and I have a lot of baking to do. Jerry, you'd best be on your way. I need to keep my sanity and balance."

"Mom, I need to borrow a wrench. I'm working on my truck, and I lost mine," Jerry said.

"Well, you'll have to look in the shed. I don't know where your dad keeps most of his tools. How did you get here?"

"I rode my bike." He opened the back door and headed for the shed. "Thanks, Mom. Tell Dad I'll bring it back tomorrow."

"Be sure to close the gate so the dog doesn't get out."

A few minutes later, I heard Jerry leave on his bike. Janet and I seriously started baking and stacking. We had a lot of work to do. We didn't speak for a while. The quiet lasted too long. I had to ask how her marriage was working; I knew she had something to tell me. We needed to talk.

"Janet, how are you and Henry getting along?"

"Mom," she said. "I knew a long time ago I made a mistake with my marriage. I have been all of these years trying to prove to friends and family that I could make it work, but I'm so unhappy. I don't love my husband anymore. He wants children. You know with my diabetes, I'm afraid to have a baby. I need more of a life than I've had. All I do is work and never get anywhere."

"Does he know how you feel?"

"He knows I'm unhappy. We have talked about it."

"Janet, are you still thinking about the guy you told me about at the Star Trek meeting?"

"Yes," she said quietly.

I was hoping it would work out for them, but in reality I knew big changes were coming into our lives. Janet wasn't her usual happy-go-lucky self. She wasn't smiling and singing. While concentrating on her baking, she was putting a lot of pressure on the rolling pin. I saw that the dough was too thin. I took the rolling pin from her, held her shoulders, and she turned to face me. We shared a warm hug, and then returned to our baking.

"Life never turns out the way we expect. I'm having the worst problem with your dad and his drinking. Then there is Danny's spirit haunting me, and I worry about you and Jerry. What are you going to do, Janet?"

"I'm going to get a divorce."

"Oh, Janet...you've made up your mind. How are you going to manage? Are you going to move? How does Henry feel?"

"I'll be okay, Mom. My job is stable and I'm looking for an apartment. Mom, if you will be strong and be there for me, then I will be there for you."

I couldn't argue with that. God knows I needed strength as much as she did. We stopped for dinner. I placed a casserole of chicken in the oven and put a pot on the stove for rice. Janet set the table.

"Your dad enjoys our family dinners on holidays, and when we get together on weekends. I've worried that he won't be with us for long."

"Mom, I can see he isn't well. If I had money, we could take him to the hospital. Life doesn't seem fair. How can I help?" She looked in my eyes.

"Thank you for wanting to help. It's hard with Christmas coming. Holidays are so stressful for everyone. If you, Henry, and Jerry can keep your bills paid up, and take care of yourselves, it is all I ask. I know your dad would be content with that. You know there were a few times when your dad and I ate canned beans and bread for dinner. He will always peel potatoes for home fries and eggs. It's his favorite dish."

"Mom, if you need food money, you know I'll help."

I sat in a chair by the kitchen table and breathed a deep sigh. I looked at my daughter and silently said, *thank you*. "We are not starving. I'm trying to get an appointment with a doctor in the Medicaid program. I know on the last few jobs your dad signed, he had problems and ran short of material. Then it's been raining, sprinklers wet the work area, or the measurements were underestimated. I want to get through New Year's. It's all I can do right now. I know Grandma's spirit is with us. I'm keeping my faith and my path."

We finished baking, cleaned the dishes, and stacked cookies to be iced the next day. Janet boxed half the cookies to take home and would ice them later. She was careful in governing her diabetes, but I knew she would enjoy a cookie or two. I was tired, so I showered and went straight to bed.

I fell asleep, and dreamed deeply. I found myself flying into the future. First, I was shown a room with a casket. Everything was a different shade of red. I had the feeling it was an emergency situation. I could not see the corpse. After seeing the casket, I was shown a wedding. I was dressed in white and silver cloth. Then I heard the words, *"Forty-five days."* I awoke with a heaviness in my solar plexus.

The following morning, I awoke later than usual. It was almost nine a.m. Alex was in the kitchen working a crossword puzzle. The coffee pot was perking. I kissed him good morning. He stood up and gave me a hug. I could feel his ribs through his white T-shirt. His hair was thinning. I glanced at the few blond strands on top of his head. When he smiled, I noticed his two loose front teeth had fallen out. I had seen Alex every day. At that moment, he looked like the walking dead. My heart was in my throat as I held back the tears. I reached for his hand and asked about his plans for the day.

"I'm going to check a job site and then stop and see a friend," he said.

I lifted a cup out of the cabinet and poured my coffee. "Okay, just be careful." I watched the front door close behind him. Then I heard the truck back out of the driveway. I smelled something burning. I found a frying pan on a hot stove burner. *Damn! Alex forgot to turn the stove burner off.* Quickly, I grabbed a hot pad, removed the pan, placed it in the sink, and turned off the stove. *Did he forget about the pan? Or maybe he didn't remember turning on the stove. Can I trust him not to burn the house down?*

I went to the bedroom to meditate. I placed my coffee cup on the nightstand and sat on our bed facing the window to the east. I felt the warm sun rays streaming through the glass.

When I finished my rosary, I did a lot of asking.

Holy Spirit, fill me with your light, healing and love. Teach me to share it with others on this planet and in this universe. Help me to make the right choices and learn for the good of all. I ask all the saints and angels to walk with me, and protect me on my journey. Bless my family and help my husband with the work he must do. I pray everyone has a safe and happy new year. Dear God, you know my needs and my wants. Your will is my will.

I look to my inner knowing for personal support. I didn't want to confide in friends.

CHAPTER **19**

Henry

We had no tree for Christmas. I set the manger on a table in the living room, surrounded by a ring of mini multicolored lights. Alex and I bought the manger at K-Mart when we were married in 1962. Thirty-six years, and there were no chips in the painted figurines. I always packed them carefully. I had won $180 on a lotto ticket, and purchased several nice T-shirts for my family.

We exchanged gifts on Christmas Eve. I gave Alex a blue T-shirt for his easy-going and calm personality. Janet received a pink T-shirt for the giving and love she shared with others.

Henry stopped by to wish us a "Merry Christmas." I handed him a present. He smiled and unwrapped a red T-shirt. It suited his dark hair and dimples. Jerry unwrapped a white T-shirt. If white is a color of protection, then Jerry, with his wild ways, could use the color white. Janet and Henry had wrapped a circular wooden plaque for me. When I opened the package, I saw an angel design burned in the wood.

"The angel is beautiful, Janet. Did you create this on your own?"

"Yes. Henry helped with burning the angel in the wood."

"Oh, thank you -- it's beautiful." I gave each of them a hug.

Alex gave me a silver charm of 69, my astrological birth sign.

"Oh, thank you, Alex. It's beautiful. How did you save the money for it?"

He clasped the silver chain behind my neck while I held my hair up. "I had some money put aside."

We had tea and cut-out cookies. I was pleased we took the time to make them. Then we exchanged hugs and we said our goodbyes.

Later that night, Alex searched through the kitchen cupboards for a beer he put away a few days earlier.

"Are you going to drink a warm beer?"

"I'm going to put it in the freezer for a while." I retired for the evening hoping Alex wouldn't fall asleep and forget the beer in the freezer.

I attended Mass alone on Christmas morning. In church I watched families as they sat together. Janet and Jerry hadn't attended Mass for several years. Alex wasn't interested in church. It didn't bother me going to Mass alone. I would have liked to have my family with me. But if they didn't want to be there, I wouldn't force them.

I arrived home from church and found Alex peeling potatoes. I was touched that he wanted to help with the dinner preparation. Janet set the table with a red tablecloth and Grandma's china. We had a quiet dinner of roast beef and mashed potatoes with our family. We missed Henry, who told me he couldn't join us for dinner. I guess it was just too emotional for him.

Jerry brought me a red poinsettia plant. I placed the poinsettia on a table beside the manger.

As the four of us sat around the dinner table, we made light conversation. Janet reminded Alex of the time he helped paint the dugout for her softball team. Jerry talked about the days when Alex was a Boy Scout leader. They were trying to help Alex find reality, and a reason to quit drinking. I reminded Alex of how he used to hide his mother's cigarette lighters.

"Your mother used to get so angry with you. Every time she wanted to light a cigarette, she had to find her lighter."

"Yes, and then she would giggle as she slipped my Bic lighter in her pocket," Alex remembered.

"Well, it's a good thing your father was easy-going. You're a lot like him."

"Yes." Alex smiled, remembering the good times.

Suddenly, Janet said, "Mom, I found an apartment close to my work. I want to move this week, before New Year's. Some of my friends are going to help me. I would like you there."

"I would love to see your new apartment. Of course I'll help."

"I took off work this week so I could get settled. I'll call you on Tuesday."

After dinner, Janet and Jerry left. Alex fell asleep on the sofa. Our dog Princess lay on the floor in front of the sofa. I heard her whimpering, and went to see why she was upset. I found Alex with a lit cigarette in his hand, and saw a brown spot had been burned on the sofa cushion. I stroked Princess on the head. "Good girl." Then I removed the cigarette and put it out. This was one of the many times I had to watch Alex. I couldn't trust him, afraid he would burn the house down. We hadn't gone anywhere for so long. Alex had always spent our last dollar on a six-pack of beer. If that was what he needed for the time he had left, I could find peace with it.

In my bed, I held a legal pad and pen, poised to write. I hoped my mother was near.

"Merry Christmas, Mom."

"Yes. I am here for you."

"Why do my thoughts conjure up so much trouble? As hard as I try to keep a positive attitude, I am always unsure of myself. The situation in my life is not improving."

"You will be okay because I will help. You will have the truth soon. I can't tell you, because you will not allow it."

"How could I stop it?"

"You couldn't stop it. You would be unhappy and that would ruin the plan."

"What do you think of Red?"

"He is in the hands of God, and is very good at what he does."

"It is hard for me to believe I will be able to work with Spirit."

"I will teach you how to use the energy. You will work with people of all kinds."

"Alex told me so many times he would quit drinking."

"He is not well."

"Suddenly I don't want to be old so fast. Do you tell me what I want to hear?"

"No. I am here to help."

"Thank you for helping. Where do you go when you are not here with me?"

"I am here to trip the men, and I go to the other side in spirit to work with those who need me."

"I wish I knew so much more."

"I am here to help. I love you. You need rest."

"Good night." After I laid my legal pad on my nightstand, I wondered what Mother meant by "trip the men." I felt it might be to head off trouble. I had to admit to myself that Danny's spirit helped me through my grief and troubles. The romance of his spirit -- and the mystery -- distracted me. It was the gift Danny gave me at the flea market. The love energy comforted me in his presence, despite his interference and trouble in my life.

On Tuesday morning, Janet drove me to her home. We made several trips back and forth to the new apartment. It had two bedrooms, and a nice kitchen and dining area. Glass sliding doors off the living room opened to a patio. I was impressed. There were no glass sliding doors or patio in Janet and Henry's home. Janet had all of her dishes, pots, and pans. Friends had given her furniture.

HENRY

I thought she was blessed to have everything fall into place. I was sure my mother's spirit had helped.

Janet and Henry had kept their disagreements to themselves. Alex and I gave counsel only when asked.

One day, Henry left his job on the construction site early and came to our home. Henry sat on the sofa in our TV room. He put his elbows on his knees, and his head in his hands. I offered him a soda, and set it on the coffee table. Then he looked at me with tears in his eyes.

"Mom, what did I do to deserve this? I love Janet. I take her blouse out of the laundry basket just to smell her. She's not being truthful with me."

I reached for his hand and looked into his liquid blue eyes. "Henry, people change. Life is not stagnant. As a couple, you and Janet were married very young, and now you have outgrown each other."

I had helped Janet pack clothes, dishes, and furniture. Now I couldn't find the words to comfort Henry.

Just then, Janet came in the front door. She stood in the doorway of our TV room and looked at Henry in his cutoff shorts and work shirt. There were tears in her eyes. Her blond hair was down, bangs covering her eyes.

They looked at each other. Then, Henry asked, "What did I do to deserve this? I love you. You're not being truthful with me. Tell me. What did I do?"

Her arms were folded, her head was lowered, and tears rolled down her cheeks. "You didn't do anything. I'm just so unhappy. I need to change my life. You want children and I don't. If I go into diabetic shock or worse, what would happen to the children? I have to work, and I want my freedom."

"You know we could adopt children. We could work this out. You're breaking my heart." Henry was earnest.

Janet stood her ground. "I work all the time and have nothing to show for it. I want a career."

LOVE FROM THE OTHER SIDE

They left our home, separately, each headed in different directions. I slumped on the sofa and wiped moisture from my eyes. I wondered if I would ever have a family again and what it was they were not telling me. I knew in time the truth would come out. Janet had always been strong-willed and independent. She put up a good front and wouldn't want me to know she had a problem.

It wasn't long before Henry also moved out of their house, and a "for sale" sign was placed in their front yard. Later, I found out Henry moved in with a female friend whom they had known for a long time. Then, I knew part of Janet's problem, and things they hadn't said.

Janet planned a New Year's Eve party in her new apartment. I was amazed how fast they each adjusted, once they decided to take different directions. I meditated and prayed each would find a new path to discover love and peace. I knew that we all would continue to grow and learn.

CHAPTER **20**

January New Year's/Yard Sale

Alex and I brought 1997 in quietly. On New Year's Eve, we watched the ball drop on TV in New York's Times Square. I took Alex's hand and pulled him off the sofa.

"Dance with me," I said.

Alex held my hand and I felt the warmth of his arms around my waist. We danced to *Auld Lang Syne*. *"Should auld acquaintance be forgot, and never brought to mind/ We'll take a cup o' kindness yet/ For Auld Lang Syne."* When we stepped to turn, he lost his balance. I laughed.

I heard Mother's spirit. "Look outside."

I opened the front door and watched a falling star, bright blue fire descending from above.

"Alex, come here," I yelled. He stepped out the front door. The falling star was gone and we watched the fireworks. Red, blue, white, and yellow fire lit up the sky. Alex finished his last beer, and we retired for the night.

On New Year's Day, I made our traditional Polish dinner with mashed potatoes, spareribs, sauerkraut, biscuits, and apple pie. Janet spent New Year's Day with her friends in her new apartment. Jerry came to the house for a home-cooked meal.

LOVE FROM THE OTHER SIDE

On the following Sunday, Alex decided to attend Red's service with me.

"I want to hear his New Year's predictions," he said.

I was elated. Alex was sober for a change, Thanks to a lot of combined effort, a ton of red tape had been cut. Weeks of communication with the Department of Health and Rehabilitative Services, Medicaid, and government disability health insurance had finally yielded results. Alex had a doctor's appointment for the first week in January. Things were actually looking up.

The weather was cool. I dressed in my usual jeans and gray sweater. Alex wore his leather jacket over his shirt and jeans. We rarely went anywhere if we had to wear dress clothes. We couldn't afford it then -- and besides, it wasn't our style. We climbed into our blue Mazda work truck, the only vehicle we had, and headed for I-95 and The Haven for Spiritual Travelers. With Alex behind the wheel, I was in the mood to talk.

"You know, I have wondered why you were never interested in Red's background or how he got started."

Alex placed a cigarette between his lips. "Look in your pocketbook. I put my cigarette lighter in there earlier. I just go along for the ride. But I know you're going to tell me. Now you've raised my curiosity."

"You have been to Red's with me many times, and you've seen how he works. Red has testimonials from people who had a healing through him. He does not ask for money, only a donation for his church. Some of the letters he's received said they did not give any money," I explained.

"If he is that good, he would be out of that old home and living in a castle."

"I'm sure Red enjoys recognition when he receives it. But he likes simplicity. He doesn't need pranksters like the teenagers who left dead chickens on his front doorstep," I said.

"You're kidding." Alex chuckled. "Someone did that?"

While searching for Alex's cigarette lighter, I found a pamphlet of Red's stuck in the side pocket of my purse. "Ha, ha," I said. "Let me read this and enlighten you." I handed Alex his lighter and watched as he lit his cigarette.

Then, I began reading, "Reverend, Edward (Red) Duke was ordained in the Universal Church of the Master, located in California, in 1970. He devotes his time to healing and guiding others along the path to higher consciousness through workshops and private counseling. His compassion comes from his own suffering. He was exposed to two atomic radiation bombs when he served in the Navy at Bikini Atoll during World War II."

"That was quite a speech from you. I know you have been going to The Haven a lot this past year. I see how accurate Red is with our readings, but how did you come up with all of this information?" Alex looked puzzled.

I looked at him and smiled. "The information is all there, on pamphlets by the table where his billet envelopes are."

Alex exited I-95. It felt so good to have him almost normal again. He parked our Mazda in the front yard. Red's flagpole, as always, was lit up. The American flag was waving in the wind. Red was standing outside. He greeted us with a warm smile and friendly handshakes. Alex stayed outside to help direct and park incoming cars. I saw a wooden plaque nailed to the front door frame. It read, "Don't let the dog out -- no matter what it tells you."

I entered the Florida room. Chairs were still in rows of five. Meditation music played from the speaker. It always relaxed me. Red's home was overcrowded with crystals, statues, and books given to him by friends who visit his home. He refused to part with any of it.

His assistant, Dawn, walked up to me and gave me a hug. I caught the aroma of lavender perfume, and she greeted me with a smile and a kiss on the cheek. I held a billet, sat down and began to write. *Could you give me some insight to solve my problem?* Over the last year, I had asked Red so many questions, but

I needed more answers. How do I solve this? My mystery was still abundant. What truth did Danny want me to know? If Red knew and couldn't tell me -- well, why not?

Alex entered the room and chose two seats in the center of the aisle. Friends began to find their chairs. Red, his red hair tied back, wearing a tan sport shirt and slacks, began his service. He stood in front of the podium with a warm smile and arms folded.

"My, oh my, look at all you folks come here for New Year's predictions. You all know I'm here every Sunday to heal and to help. This crowd is like church on Christmas."

"Hello, Red." The crowd responded with chuckles.

"Well, let's see how in tune you can sing tonight."

Dawn passed out music sheets. The spiritual friends started to sing and raise the vibrations in the room. Red walked through the aisle, shaking hands and greeting guests.

Reverend Edward (Red) Duke began his lesson. "The most common technique of healing is as old as civilization. The healing practitioner or psychic uses the energy within his or her own nervous system. This energy is concentrated upon to intensify its healing qualities, and is then transferred through their palms or fingertips to the patient. The healer's hands are placed over the congested area, making physical contact or contacting a few inches above the body, or touching the head of the patient, sending the energy to the diseased area. Healing can occur gradually or instantaneously.

"I received my gift of healing after surviving radiation poisoning while I was in the Navy. I believe nothing happens by accident. God gave me this gift for a reason. I will help anyone who asks me for a healing. For friends who ask for a healing and do not receive it, nothing is lost and something is always gained. Some of the methods used in spiritual healing are:

"Body Balancing – this puts one in a state of harmony by aligning the spiritual, physical, and mental bodies.

"Shrouding – saturating the body with spiritual energy.

JANUARY NEW YEAR'S/YARD SALE

"Remote Healing -- Mental healing or visualization to guide a projection of spiritual energy, used when a person is out of reach or otherwise untouchable. Please feel free to ask additional questions and view some of my testimonials."

Red then read from the Bible. We had a meditation, and he read a benediction. Then the crowd took a break.

Inhaling the fragrance of jasmine, I had just sat on the backyard swing when I heard Red's cowbell. I strolled from the back to the front of the house, where Alex was having a cigarette. We followed other friends through the door into the Florida room. Red was waiting at the far end of the room. He placed his cowbell on a wall hook, and reached for the billet box. Several billets were read before mine. Then he looked at me.

I knew he liked people to respond. I held my hand up. "Hello, Red."

Red lowered his head in thought while holding my billet. Then he smiled. "I'm hearing the name Betty. Do you know a Betty?"

"No, I don't recall a Betty," I answered.

"Say, not yet."

"Not yet, Red."

Red continued, "There is a spirit you gave light to on the other side. His chest was all burned. He wants to give you rubies and diamonds."

Red looked at Alex and chuckled, "And I know you want to know when."

I remembered the dream. It was almost a year ago. In many of my dreams, I felt out-of-body travel and energy with other spirits. How did Red come up with all his information?

"Let me come into this vibration?" Red asked, holding my billet.

"Yes, Red," I answered.

"You're getting more relaxed," he said, alluding to my meditation. "Your destiny will be shown soon. It will happen and you will have no control over it."

145

"Why couldn't I change the energy? Aren't we in control of our actions?" I asked.

"When you control situations, you lose the lesson the spirit side has for you. Be patient. Accept fate without control. The universe will fill the need."

"Thank you, Red."

We were nearing the end of the evening, when Red said, "I saved the best for last: predictions. It's going to be sunny and hot in southern Florida this year."

The crowd chuckled.

"I don't have strong feelings about storms this year. Don't retire your hurricane shutters, but for this year, it looks pretty good. I feel England in the news. I have a strong feeling for the royal family. You all are rising to a higher frequency and vibration."

The crowd disbanded, hugs were exchanged, and we said our farewells.

On our drive home, I told Alex that Janet wanted to have a yard sale. I knew it was only a matter of time until the bank gave us an eviction notice. We just could not come up with four months of back mortgage payments, and they wanted it all at once.

"You know, when we have to move I don't know what I'm going to do with all of our dishes and presents we have collected through the years. We could combine our stuff with Janet's and have a big yard sale," I said.

"I'll help. We sure could use the extra dollars. I'm sorry things are in such a mess. I can't find my way out of this. I can't believe our daughter is getting a divorce." He was distressed.

I reached across the front seat and touched his arm.

We arrived home around 10:30 p.m. Alex opened the fridge for a beer. Our dog was waiting by the front door, so Alex went outside for fresh air. Princess followed for some exercise. I didn't have a chance that evening to take my walk, so instead, I chose to do a few sit-ups on the bedroom floor. It wasn't long before I

JANUARY NEW YEAR'S/YARD SALE

heard the front door open and close. Alex had finished his beer and entered the bedroom.

He looked at me exercising on the floor and said, "What does that do, tighten up your pussy?"

This certainly was not Alex. He had never used dirty language around me. I knew Danny's spirit had walked into Alex. I figured Danny's spirit was trying to cause conflict between us. I knew that earthbound spirits absorb energy from negative occurrences. I looked at my husband. He had negative energy bouncing off him.

"What do you care? You like assholes," I said. I felt his frustration as he left the room.

Just then, while I lay on the floor, Princess came over and licked my face. I felt her warm wet tongue on my cheek and I had to laugh. I put my arms around her neck, and hugged her. She helped me keep my inner peace.

The following week, our whole family worked for a yard sale. We bought bright green poster board and painted signs with arrows and times. Janet brought over boxes. We spent a day sitting at the kitchen table unpacking and pricing items. My daughter and her husband had divided their possessions. I saw their hurt expressions and at times, tears in their eyes, as they separated and decided what each would keep. Henry wanted his Indian bows and arrows. Janet wanted her unicorns, and they divided pictures autographed by *Star Trek* actors.

Most of their friends had expressed shock at the breakup. The general feeling was that out of everyone in their circle, they thought those two would be the ones to stay together.

Alex and I didn't have much of value. Our furniture was second-hand. I knew we couldn't charge a lot of money. Unpacking boxes to eliminate things I had kept, I carefully removed the paper from a statue of an old man and woman dancing. I turned the key on the bottom and it played "The Anniversary Waltz." My parents were there in spirit. I looked at Janet sitting across from me at the kitchen table.

LOVE FROM THE OTHER SIDE

"I gave this statue to Grandma and Grandpa for their 50th wedding anniversary."

"I remember that statue. I thought we would be married for 50 years," she said.

I watched Janet glance at her wedding picture; she had placed it earlier in the day on the kitchen counter.

I placed the statue on the table. "Nobody knows what life has in store for them."

We listened as the music gradually slowed, then stopped. "I can't bring myself to part with it," I said, the moisture building in my eyes. "I'll keep your wedding picture. It holds memories of a time in our life when there was joy."

Carefully I rewrapped the statue, and then put aside glass vases, wedding presents to Alex and me. We went through boxes, unpacking and repacking items and memories.

It was the middle of January 1997. We sat in front of our home under a blue tarp to keep out of the hot Florida sun. Several friends and neighbors brought over items to add to our collection. We had T-shirts, furniture, candleholders, dishes, and a few tools from the shed -- and some junk. Henry had decided not to join us.

One gentleman picked up a video off the table, looked at me and said, "I'll give you two dollars for it."

"Two dollars! I'm asking five."

"It's an old movie. Everyone has seen it," he said.

"*Gone With the Wind* is a classic. I paid twenty dollars for it. I want five. Take it or leave it."

He put it down on the table. "Better luck with your next customer."

He left. There was a man who looked to be about sixty years of age. He sauntered toward our table. His gray beard was scruffy and his T-shirt, well-worn. He had a wicked, repulsive energy around him.

"Do you have any ceremonial swords?" he smiled, showing a few missing front teeth.

"No!" I said defiantly. He had caught me off guard and I was angry.

"What was that about, Mom?"

"I don't know, Janet. I've had some weird experiences this past year. They are always spontaneous. I have learned to brush them off or ignore them."

A blond-haired girl looked to be about six years old. She was admiring a bracelet with pink stones on a table. I watched her pick it up and try it on. She looked at her mother with anticipation. "You don't need that," her mother said.

The young girl looked sad, as she walked around looking at sale items.

Janet walked up to the mother and said, "I'll sell it to you for one dollar."

I heard the mother say, "Okay."

The child smiled excitedly.

The sun was hot. There was barely a light breeze. A tall, thin man in a long black coat, wearing a matching fedora hat, walked into our yard. He looked over items we had on the tables and then said, "Tell you women what to do and you don't do it anyway." Then, he left.

"That was weird. I wonder what he was talking about," I said to Janet.

Neighbors stopped to shop and to thank us for our Christmas cookies. Our yard sale was a success and we made a little over four hundred dollars. We ordered pizza. Jerry drank a beer with his dad. We sat around the kitchen table after our hot day in the yard and I told the family, "Dad had a doctor's appointment two weeks ago. We're waiting to hear the results. This Monday, I have a job interview. All of those applications I filled out paid off."

"Mom, that's good news. I sure hope you get hired," Janet said.

"Janet, can I borrow your car for my interview?" I asked.

"Sure, Mom. Anything I can do to help."

"I would take you, Mom, but my truck still isn't running just right."

"It's okay, Jerry, and quit feeding Princess your pizza crust."

Alex stood up and gulped the last of his beer. "Jerry, let's finish putting the tarps away and sweep the driveway."

"Sure, Dad."

Janet helped clear the table and we hugged each other. I closed the front door behind her and wondered what my new job would be like.

CHAPTER **21**

Crossing Over

That week, Alex drank beer continuously, morning and evening, taking naps throughout the day. There was no talking to him. He was not coherent. He turned the stove on and then forgot about it. He looked through cabinets for beer cans. One evening, Alex came into the bedroom after I had an active day. He wanted sex. I refused him. He was drunk, really wasted.

He started in on me. "Treats us like ca-ca. Won't give us ten minutes of sex. After thirty-six years, this marriage is over."

He slept on the sofa in the TV room all night. After he had left the bedroom, I released a sigh of relief. "Peace," I said to myself.

Dear God, I need help. I reached for my pen and note pad on the nightstand.

"Hello, Mom?" I wrote.

"Yes, you are loved by all. Danny will be sorry he messed with you."

"Mom, do you have another name in Spirit?"

"Yes, it is in the plan and you will know it soon. But for now, I am Mom."

"Tell me something about Danny that he does not know."

"He is in the way of me and not the way of me. Time. He has

not the time in his plan for what he wants, so he will have to lose it."

"I thought in my dream I was supposed to help Danny?"

"He is a troubled spirit and we will help him. But you will be the key, and he will change."

"Mom, I loved Danny, and I don't want him to suffer. Will he find peace? And I don't know where we will be living, or if I will be hired for a job. I'm so confused. My faith is all I have right now."

"You are the key and we will help him. You must get some rest. You will be very busy. Things will change very fast."

"Thank you. I love all of you in Spirit and on the earth plane."

"So many times, we come by the way of love only to lose it."

"Good night."

The following morning, I awoke to the aroma of coffee drifting in from the kitchen. I slipped into my robe. My taste buds anticipated a hot cup of coffee. I found Alex sitting by the kitchen table staring at a crossword puzzle.

"What's another word for salt?" Alex asked.

"Saline."

"No, I need another word, a synonym."

"Look up the answer."

"Okay." He looked away from me; his eyes were cast downward.

I sat next to him at the table, poured coffee in my cup and asked him, "Do you remember last night?"

He was still wearing the same T-shirt and cargo shorts. He looked distressed.

"I'm not sure. It was like standing outside of my body and watching a movie."

"I know it wasn't you. In all the years we have been married, you never treated me like that." I bent over and kissed his cheek.

"I'm so sorry," he apologized.

I noticed a tear in his eye. "What am I to do with you?" I said.

CROSSING OVER

Dear God, I knew Alex at times had lost control and given in to spirit energy. His ego must be mutilated -- no work or money, the bank foreclosing on our home, and his health deteriorated. Of course drinking beer wasn't the best medicine. But with our situation, as bad as things were, if beer took the edge off his tension, I wasn't going to argue with it.

"I have to see someone this afternoon. Things will get better. I promise." Alex was earnest.

"It's that man who has been loaning you money, isn't it?" I asked.

"Yes; after we tiled his pool deck, he liked the result. He has been a big help to us," Alex said.

"You're right. Without him, we would have starved," I said.

"Well, if shit were worth something, a poor man wouldn't have an asshole," he said in defiance.

Alex stood up, and placed his coffee cup in the kitchen sink. I gave him a hug. He put his arms around my waist and said, "I love you."

"More!" I said, with my head against his chest, I could feel his heart beat. I wasn't sure how, but I knew I was going to lose Alex.

He looked me in the eyes, kissed me on the lips, and turned away.

I watched his six-foot-tall thin frame of fifty-nine years. His shoulders were bent as he walked toward the bathroom to wash up and change clothes. I knew he was depressed, but I hoped he was determined to keeping trying to survive the days. I left to meditate in the bedroom.

Our voices echoed in the nearly empty house now. After our yard sale, we had cleaned out all but our basic furniture. We were left with the bed, dresser, kitchen table, one sofa, and our TV. I heard Alex back our truck out of the driveway. I slid the drawer out from my nightstand, and reached inside for my rosary.

After my meditation, I fell into a deep sleep. In my dream, I was walking. Then I stepped down, and immediately entered a bright white light. I suddenly came awake, startled.

What now? I had listened to people speak of a near death experience. They mentioned traveling through a tunnel and at the end finding a bright white light. For me there was no tunnel, just a bright white light. I made a note on the legal pad that lay on the nightstand beside my bed. I continued with my daily chores.

Putting my dream and thoughts aside, I washed breakfast dishes and removed hamburger from the freezer for dinner. One reason I kept notes of my dreams was so I could put my messages aside and continue with my daily life. If I worried about or dwelled on my premonitions, I would be reading ink blots in a doctor's office. I was startled by the ringing of the telephone.

"Hello."

"Hi Mom – it's Jerry. Can I come over for dinner tonight? I just want to be with you and Dad."

"Sure Jerry, you know you are always welcome. We're having hamburgers and french fries. You can watch a movie with Dad."

Jerry rode his bike over in the evening for dinner and to watch television with us. The date was Wednesday January 22, 1997. Alex was feeling the effects of his beer. I had just finished washing dinner dishes when he came in the kitchen and put his arms around me. He said, "I am going to walk to the store for a pack of cigarettes."

Then he said to me. "Carol, you have no guilt. What I have done, I have done to myself." I heard the front door close.

After a short time, Jerry and I noticed Princess sitting by the front door whimpering. Jerry told me, "Dad has been gone too long. I'm going to look for him."

He hopped on his bike and left. I became frightened, my heart was in my throat, and I ran out the door to follow him. *Oh God! It was the light I fell into after my meditation this morning.* It was dark outside. I ran toward the main road Alex had to cross to get to the Circle K. There were several police cars parked on the far side of the road. I saw Jerry standing there talking to an officer, and he called to me. Blue lights flashed in my direction. It seemed forever until I reached the other side of the road.

My son threw his arms around me and sobbed.

"HE'S GONE!" Jerry cried. We hugged each other and cried together. I noticed my husband's Dolphins cap lying on the hood of the police car. "They said he walked in front of the car," Jerry said.

My first thought was that Alex did it on purpose. My second thought was that his mind was somewhere else. He didn't know what hit him.

"Jerry, go call Janet and Henry. I don't want Janet to drive here after you tell her about the accident." I watched Jerry cross the road on his way to our house.

A young woman walked up to me and asked if she could give me a hug. She motioned toward the group she was with and told me, "My friends and I just left a church service. We had the overwhelming feeling to take this road. We prayed for him before they took the body away."

She asked if they could pray with me.

"Thank you for praying for my husband," I said, wiping the tears from my eyes with a tissue. "But I would like to grieve in private."

Janet and Henry joined Jerry and me. We all hugged each other, and the tears flowed. Deep sobs punctuated our sadness. I watched Janet and Henry hug each other, for the first time since they separated. I knew they would not get back together. But in that moment, I felt a healing wave of energy pass through me. The officer handed me Alex's Miami Dolphins cap. It was the only thing I had left of him.

"Your husband had no ID on him. If your son hadn't shown up we wouldn't have been able to identify him," the officer told me.

I asked my daughter to call Red, and our family in Ohio. I didn't think I could handle the surges of emotion.

CHAPTER **22**

Letting Go

Later I was alone in the house. I had asked Janet to go to her apartment and telephone my sisters, family, and friends. I knew, with tears flowing, it would be hard on her to spread the news about her father. But I just couldn't do it. I told Jerry to go home and call his boss in the morning for time off work. He needed to call his friends and put his apartment in order, because we would be very busy for the next few days. I didn't mind being alone. I wanted to grieve in private and make my peace with Alex's spirit. In bed, I laid Alex's Dolphins cap on the pillow beside me. Princess slept on the floor by the foot of our bed. She gave a sigh as her head rested on her front paws. The toy that was always between her paws was not there. I could feel sadness in Princess, the sorrow in her heart. I cried deep heaving sobs, and finally slept on my pillow, which soon was wet from tears.

The following morning, just before I awoke, Princess let out a long, high-pitched howl, the most sorrowful howl I had ever heard. Alex came with the light I had seen just before Princess howled. The light started to leave. I said "MORE." Alex stayed a few seconds longer and I felt all tingly in my solar plexus. It was my moment of profound insight, a glimpse of love from the other side, my Epiphany. I glanced at his hat that lay on the pillow beside me. I knew Alex was in the light. His light was a ray of warm bright sunshine on a gray winter day.

The following day our dog lay on the sofa, where she used to place her head on Alex's lap. I opened the front door to pick up the afternoon mail and found potted purple mums on our front doorstep with a sympathy card. I didn't know if the officer had cited the driver with a ticket. I never had a chance to meet the couple from Alex's accident. My eyes filled with tears. When I opened the mail I discovered a letter from the Department of Health and Rehabilitative Services. It read, "We have reviewed you medical reports from our physician. It has been determined you are not eligible for benefits. If you think this action is incorrect---------------."

My chest felt heavy. Alex must have known he was not going to receive help.

The next few days were like a bad dream. I needed to get death certificates and an accident report. There were no customers to call; Alex had not signed any new contracts. Jerry, Janet, and I made arrangements to have Alex's body sent from the county coroner's office to a cremation society. I never saw his body. My sister Emily and her husband Fred flew to Florida from Ohio, to attend the church service for Alex. Fred was a retired fireman. My eldest sister Vivian telephoned and informed me she and Larry couldn't make it. They gave their sympathy and condolences. Jerry was animated as he mopped the floors and cleaned our house. I knew he wanted and needed emotionally to help. He kept shouting at Princess to get out of his way, and then he finally chased her out the back door. Later that afternoon, Jerry gave Princess a bath on the back patio.

"Thank you Jerry for helping."

"It's okay, Mom, I want to help. If I had gone to the store for Dad, he would still be here."

"Jerry, do not feel guilty. Incidents happen that we have no control over. It's not your fault; Dad chose his time and place, his time to leave us, Jerry. He is at peace now." We hugged each other and shed our tears.

My mother's spirit sent me a mental message, *"Play a polka to remember me."*

I knew she was with me. Mother's spirit had walked with me through all the poverty and tragedy I had been dealing with. I knew what she was telling me. Mom was sending me a message for a hopeful and brighter future.

The Friday after Alex's accident, I had a call for a job interview.

"Hello, Carol Shimp?"

"Yes, how can I help you?"

"I'm calling from Publix. I have reviewed your application for employment. Could you come in for an interview on Monday morning?"

"Oh yes! I'm so excited. I really need the job." Under my breath I said, "Thank you, Mother."

"Can you be here at ten a.m.?"

"That will be perfect."

"Just ask for Tammy. See you on Monday."

I met with Tammy, the hiring assistant at a neighborhood Publix, our local grocery store. She asked if I could start training for a cashier's position that week. I was so relieved that I could hardly breathe. I told her yes. But when I informed her of my husband's accident, she grew disturbed.

"How can you start work when you're going through an emotional crisis? If you need more time, you can have it." Tammy looked concerned.

"I'll be fine." I said. "I need to work. It will be good for me to keep my mind busy." Tammy pulled two uniforms off a shelf and handed them to me. There were two green vests and two pairs of navy slacks with two white cotton blouses. While training for work I had moments that week when I had a hard time holding back my tears. Keeping my mind busy and learning new procedures helped me get through the day.

We had a Mass said at St. Clements, the Catholic church I had been attending. It was scheduled a week after I started work. I asked for the day off. I found a pair of tan slacks and cream blouse in the back of my closet. Although they were slightly large on me, I made the outfit fit with a tuck here and there. My children and I, along with my son's girlfriend, piled in Janet's Camaro and headed for the church. We hadn't yet received Alex's ashes, but I handed the priest Alex's hat. It was all I had. The priest placed the orange and teal cap on a podium in front of the altar before Mass. While wiping my own tears, I heard quiet weeping and blowing of noses from friends, employees, and family throughout the church. None of Alex's family could make it to the church or our home. The priest eulogized Alex. He held my husband's cap, rotating it in his hands.

I watched as our priest held Alex's cap. I'll be forever wondering whether Alex walked in front of that truck on purpose, or if he just didn't know what hit him.

"I did 'not have a chance to know this spirit. I'm told he worked hard. He took care of his family with love and commitment. Alex Shimp extended service to help those who asked for it. I am told in his more affluent days, he bowled a 300 game. He didn't boast about it. Now I will share a tribute, by his loving wife, Carol."

> Dear Alex,
> God has welcomed you home.
> He spread his arms,
> Surrounded and embraced you with warm, bright, loving light.
> You have left us with our memories and the essence of your love.
> I'll miss the warmth of your arms around me,

LETTING GO

And in the morning, the aroma of coffee drifting into the bedroom.
Thank you, Alex, for loving me, and for taking care of me.
Thank you for Jerry and Janet.
Thank you for the songs you sang, and for holding me in your arms while we danced.
Thank you for the hard times, the good times, and the laughter.
I pray as you stand in God's crystal reflection, your arms will spread wide, and you will embrace and surround me with warm, loving, bright light when I come home again.
Love, Carol

After church, we all gathered in our home. Food and money came from friends.

I could not believe the abundance. I had little of my own money to draw from. Jerry kept Princess in his apartment so she wouldn't be disturbed by all the people. My sister and brother-in-law stayed for a few days. She gave me strength to help me through my grief. We went to dinner and shopping. Janet, Jerry, and I relieved stress by relaxing in the hotel swimming pool where Fred and Emily were staying. Emily's husband Fred swam laps in the pool. In the spa, I slipped into the warm water and was immediately put at ease. I enjoyed the reunion with my sister and my children. Emily told me there was plenty of security at the hotel, but the fire alarms kept going off all night.

"Danny, my troubled spirit, must have triggered the fire alarms. Because Fred is a retired fireman, Danny probably got a chuckle out of it."

"It was probably some kids in the neighborhood," Emily said.

"Really! Did security catch the children?"

"No, they couldn't find them."

"Well, after my hang-up phone calls, getting zapped, the charge I received at the flea market, and thoughts I received from him -- Danny has the power. It's amazing what spirit energy is capable of. By the way, thanks for the garlic you sent me. I just need to let you know the garlic didn't work. You had your spirits mixed up. After all, garlic is for vampires, not troubled spirits."

"I can't believe a spirit could create all of the incidents that you tell me about. You need to get your life together," Emily said.

"I remember a story Mother told me. When our oldest sister Vivian was a small child, in the '30s, Mother said every Valentine's Day Vivian would go into convulsions. Mother told me there was an English nurse who lived in an apartment across the road from us. The English nurse told Mother the next time she had a convulsion to take off Vivian's nightgown and throw it in the furnace. We had that huge coal furnace in the basement. When Mom opened the furnace door and threw Vivian's nightgown in the fire, it flew up like an entity was inside it. After that, she never had another convulsion."

"Mom, are you saying Aunt Vivian was possessed?" Janet said with surprise. "I never heard that story."

"Okay," Jerry said with a smile. "I really have a bunch of Fruit Loops for a family."

"I recall our evenings on the front porch and all of those superstitious stories that were told. But Carol, you really need to let go of this spirit." Emily looked at me with her usual big sister's authoritative expression.

"I'm trying to. I started my new job. Things will work out for me. You know, I've been in the water so long the skin on my hands is starting to wrinkle. It's time for us to go."

We climbed out of the spa and dried off. I watched Janet gather her car keys, and Jerry as he dried off.

"Janet and I will pick you up in the morning for the ride to the airport. Thank you for the dinner, and the day out. Swimming was really a treat. I feel so relaxed afterwards."

We shared hugs, and Emily gathered her towels. I watched her and Fred walk toward the lobby and then the elevator. My sister helped me in many ways, but she was my older sister. I always felt I had to defend myself in her company.

Janet and I drove Emily and Fred to the airport for their return flight to Ohio. I knew that when the time was right, I would be going to Ohio to review a part of my past.

February

Alex's clothes still hung in the closet. While I was meditating one morning, Alex's leather jacket fell off the hanger onto the closet floor. I had kept a few things of his to give Jerry and Janet, although Alex did not own any particularly valuable treasures. I wrote the date his leather jacket fell, on the note pad that laid on my nightstand. When Janet and I picked up Alex's ashes from the cremation center, the mortician handed me a standard cardboard shoebox. The crematory had offered to sell us an urn. I wasn't happy with the shoebox, but it was all I could manage.

Before going to sleep that night, I leafed through my note pad on the nightstand. In my inner need to keep and date notes, I often referred to messages to verify the information. I noticed Alex's jacket had fallen off the hanger on the same date he was cremated. I placed the shoebox, which contained his ashes, on the closet shelf until arrangements were made to bury them.

I missed Alex and still waited, thinking he was going to walk through the front door. I wanted to hear him sing in the shower. I wanted to dance with him; I wanted to feel his arms around me gliding across the dance floor going round and round in circles. I called a veterans organization to see about a military plot for Alex's ashes. We lived on the east coast of Florida, but we had to take Alex's ashes to Bay Pines Military Cemetery near Tampa on the West Coast.

CHAPTER **23**

Publix

It was the beginning of March, 1997. There was so much to learn at work: the procedures for government food stamps and WIC, checks, coupons, credit cards, savings, and personal checks. In the checkout line, I reviewed signatures. Time seemed to stand still when customers were always in a hurry. Fortunately, with time and patience, work was becoming routine. I learned codes for grocery items and became familiar with the aisles of merchandise.

I was driving our truck to work with a "For Sale" sign in the window. I was hoping I could sell it so I could use the money to look for an apartment within walking distance to work. The house was dark and empty when I came home each night. Princess greeted me happily. When I stooped down to pet her, she licked my face. Then, she would find her leash and bring it to me.

We were into the second week of March when the bank took our home. I had no control over the house or the bank. I sold the work truck for seven hundred dollars. That blue Mazda was pretty beat up. I also sold my wedding rings. My diamond wasn't large, one quarter of a carat, but I received three hundred dollars. It wasn't a lot of money, considering the emotional and sentimental value the rings held for me.

Selling them helped pay rent, and the deposit for a small apartment I found within walking distance to work. It was a white two-story building with eight apartments. I had one on the first floor, and was allowed to bring Princess. My kitchen table was too large for the tiny dining area in the apartment. I gave it away and found a small round one with a glass top. The complex manager offered me a coffee table for my tiny living room. I thanked her gratefully. The windows had vertical blinds, so I didn't need new curtains. Things just seemed to fall into place. Jerry and Henry helped me move what little furniture I had. While sorting through boxes, I came across Mark's letter from the past November. There was so much truth in Mark's advice and information. My Dharma (law of the universe), soul path, security issues, and positive changes did come to pass. After Alex's death came my new job and apartment. Following my path, my advice from my spirit guides, who I call Mom, the universe's divine plan, and my inner knowing followed through to know what was right and at what time.

I was getting to know my co-workers and customers in the store. I had no interest in finding a relationship, but I didn't realize there were so many lonely bachelors who did their own shopping. They tried to mix me up when I was counting money. Janet called it "the subtle flirt."

I was also receiving messages from customers that were out of context. A male customer would ask me, "Are you feeling lonely and unwanted?" or "One day you're somebody, and the next day you're nobody."

One day, a young male customer laid a card on my counter and just walked away. It read "On your wedding day." I figured Danny's spirit was walking into customers. The change of energy and frequency was amazing.

I thought the wedding card left on my register was just too intense. After work that evening, I walked with Princess around the apartment complex.

"That is a beautiful German Shepherd," neighbors commented.

"Thank you. Her name is Princess."

I saw glitter, multi-colored hearts scattered on the side of the street. I couldn't help smiling.

I still could not figure out what truth Danny wanted me to know. I felt Danny's energy. I didn't know what little girl might have dropped the tiny glitter hearts, but spirit energy was there to grab my attention. His energy brushed my cheek. Like a silk scarf, it moved from my cheek to my hand. Princess stopped for a moment and looked at me with her large brown eyes. We started to walk again. I sent a thought, "I missed you for a long time."

I heard his voice. "I missed you. Let's try again."

"I cannot feel your kisses or your touch. You are not here. You're there."

I heard his thought. "I walk beside you. You will be all right."

I returned to my tiny apartment. Princess lay in her favorite place by the front window. I showered, and ate a chicken TV dinner. When I was relaxed, I picked up my yellow legal tablet, held a pen in my hand, and started my automatic handwriting. I had not picked up pen and paper since last January.

After Alex died, I had just drifted through the days. My time, it seemed, was in a void. I was so busy moving to my apartment and training at my new job. Oh yes, my spirits were still around. Danny surely made his presence known.

Now, I needed to communicate. I was still looking for answers. My mother's spirit helped me through my heartache and our loss of Alex. Her spirit gave me comfort. The energy and mystery of Danny's spirit gave me strength to walk through my days. Knowing I still needed help and to find balance in my life, I knew I needed to turn to my spirit guides for the answers.

I took a pad and wrote, "Hello?"

"You know I love you."

"Why are you still here, Danny?"

"You know I won't leave, because I want to marry you. I love you."

"We have been through this already. You need to go to the light."

"Yes, but this time it will work. Because someday, we will be together and I will tell you the truth."

"Yes I know. I can trust you, and you will change. We have been through this already."

"You know I will take care of you, and you will be happy."

"You make me crazy. You're turning my hair gray."

"I did what I had to do, and I still love you. You will be okay."

"I know that."

Did Danny have to get married? I still wondered. *Was it a shotgun wedding?*

"Me, I am Mom."

"Mom? How did you sneak in here?" I was surprised to feel Mother's presence.

"Yes. I love you. You will be very busy because things will move quickly."

"But Mom, I have already been through so many changes. I love you and I miss our card games."

"Yes, we miss you too. We don't have cards on this side."

"Thank you for helping me."

"Yes. I will be here for you. You have a big surprise coming. It's a good one. Be ready for it."

"Yes, I will accept it. Mom, has my prophecy changed? Am I following the right path?"

"Yes, you are on the right path. Your prophecy has not changed."

"I'm glad we had this time together. Tell Alex I'm sorry and I love him."

"Yes, I will do that."

"Do you have a message for me before I sign off?"

"Yes, we love you. You will be okay, because we will help you. And Danny will pay for your trouble. But, he will still love you. He just doesn't know any other way. He must learn other ways."

"I'm tired, Mom. I have to be at work in the morning."

"*Yes. Rest and be at peace tonight. We all love you in the spirit world and on the earth plane.*"

"Love and Light. Good night." I felt like no one would ever know the two different lives I live. Why should I care what anyone thinks or knows of my spiritual side? Because some don't believe, it does not change my experience. For someone who shares the experience, I pray my story will give them insight.

CHAPTER **24**

The Gift

APRIL

The church would be giving out palms next Sunday. This was the first Easter season that I hadn't planned a dinner for the family. At work, I was constantly scanning packages of jellybeans, chocolate bunnies, and Easter eggs in my checkout line. I couldn't believe it was April already. My life had changed so fast.

    I wanted to spoil myself, so I borrowed Janet's car for a trip to the mall. I had my hair cut and permed. I purchased makeup, a mirror and incense. In the bookstore, I bought "Sex for Dummies." After being married for thirty-six years, I didn't know what kind of men were out in the field. I didn't feel I would be comfortable with dating. I still missed Alex and knew I would always miss his hugs. The changes in my life had been fast-paced but I had to keep going forward. I hadn't even been to The Haven for Spiritual Travelers since Alex had his accident. I was working on weekends and time was short. I asked my spirit guides to see Alex again.

    On this morning, as I slept, I was blessed with the gift of Alex's presence. I felt the love energy as his arms embraced me. Alex came in a yellow and green light. Princess let out her sorrowful, high-pitched howl, and I knew Alex was with me. I became more alert, and he started to leave. When I said "MORE," his spirit stayed for a while longer. Alex felt young and strong like he did when we were first married. What a wonderful feeling of pure love from the other side. Later while checking groceries at my register, I felt the vibration of his energy; several times I had to hold back the tears.

The last week of April, Henry drove our family to Tampa in his 1990 white Chevy truck to bury Alex's ashes. Janet didn't think her car would make the trip. Jerry was still working on his truck after his accident. We were a family that held peaceful relationships. Janet and Henry came to compatible terms after they resolved to divorce.

Jerry sat in front with Henry. Janet and I sat in the back bed. We had a blue air mattress and a cooler with lunch, drinks, and snacks. We passed food and sodas through the back truck window as the guys asked for them.

"Looking back, Janet, I remember a story Dad told me."

"Mom, Dad never talked about his service days. I want to hear it."

"While your dad was in the Army he was a military policeman. He told me he was in a bar one evening when two sailors started a fight over a woman. He tried to break up the fight, but with two against one they left him standing there and left for their ship. Well, needless to say, Dad was pissed. He went on the ship after the sailors. When Dad tried to arrest the sailors, they threw him overboard."

"Mom, that sounds like something Dad would do. But he could have been hurt."

"Dad said he fell in the water between the ship and the dock. He then called for back-up. It's what he should have done in the first place." I glanced at the shoebox on the truck bed beside me. Alex's Dolphins cap lay beside the box. "We miss so much in life when we don't take time to share our experiences."

"I don't know why we are so busy today with work, paying bills, and keeping friendships. I feel like I missed a part of Dad and now I'll never have the chance to know," Janet said.

THE GIFT

"Life goes on, Janet -- we have to move forward. Dad loved you and Jerry; it's the love of his memory we will keep close."

Henry turned into Bay Pines Military cemetery and drove onto a paved road. The lawns were green and well-manicured, dotted with row after row of small flat rectangles of bronze and concrete gravestones. A large granite memorial rose from the center of the cemetery. Henry parked by the red brick office building. We stepped into the office to speak with the receptionist sitting at her desk. She took our names and pulled paperwork out of a file cabinet for me to fill out. Once completed, she paged a caretaker and we stepped outside to meet a workman with a shovel. We followed him, walking on the green lawn until he stood by the base of a large maple tree. He started to dig a hole. I held the shoebox with Alex's ashes in my arms. Handing Alex's ashes to the caretaker, I thought of Alex's Dolphins cap, but I had left it in the truck. I was not sure if I wanted to bury it with his ashes. A mental haze surrounded me. I was just going through the motions; my mind drifted far away into emptiness. The four of us stood there, each deep in our own thoughts. I did not have a place beside Alex and felt heaviness in my solar plexus, a lump in my throat. These were my last fragments of Alex. But Alex was not in this shoebox. He was not here. Alex was in that beautiful light that enveloped him when I felt his spirit and his arms around me.

Suddenly, I asked the workman to wait. I retrieved my husband's cap from the truck and handed it to the workman. He laid it in the earth and continued to fill the hole with dirt. I felt a release, as the ashes and hat were covered. I knew I had to let go. If I had held on to Alex's hat, neither one of us would be free.

I noticed some of the graves had flowers. In my grief, I didn't think to bring any. I wished I had. We ate lunch in the truck, walked around the cemetery, and then headed home. I knew it would be a long time before I could get back to Tampa again, but Alex had a place of remembrance and I had a sense of completion.

May

It was the first week in May. I went to divorce court with Janet in the morning. There was a long line. Her divorce was a mutual agreement and uncontested. I was surprised how quick everything went. A lady judge signed the paperwork. The divorce was finalized.

"Will you drive, Mom? After court I feel a little dizzy." She handed me the car keys.

"It's close to lunch -- we need to stop and eat. Your blood sugar must be low." We stopped at Wendy's for a chicken sandwich and iced tea. Finally, we were on our way home. Janet put her hands over her face and let out a small scream. She looked at me.

"Mom, I'm a single woman again!"

I never believed I would lose Alex, at least not so soon. The reality of being by myself was still new to me.

"You know what?" I said. "So am I!"

We let out a sigh and then a giggle. "The way this year has started I can't help but wonder what will happen next. My spirit guide told me my life would be changing fast and I would be very busy. WOW! Was that right."

Janet and I had been sharing her car. Henry had given me a ride when I needed some things. They were trying to figure out what to do about their house. It hadn't sold yet.

We are all trying to reach a higher plane. The more lessons we learn, the greater our Karma is diminished. Thinking about Karma, I decided to write my nephew Mark Dodich. I needed to update my astrology forecast and get spiritual support.

May 13, 1997,
Dear Mark,

So much has happened since your last letter. My husband passed away, my daughter was divorced, and I moved out of our home into an apartment. I was blessed to be able to find a place that would accept our dog. I'm also working at a local supermarket. It has been almost ten years since I have worked outside the home. Now I'm thrown into the reality of the real world, because I have a need for security and a roof over our heads. I'm finding new relationships and have made friends in my workplace.

Any spiritual guidance or information on planetary forces would be a great help to me get through my transition.

*Love and Light,
Carol*

Dear Carol,

Nice to get your letter, and hear that you are settling into a new place. I have lived alone for so long. It could be difficult to adjust to someone else being around. I expect when I do find that long-term relationship, we shall have to find a big place with plenty of space.

1997 actually shows more promise for material expansion. Jupiter is the planet of gain, so there is potential for improvement when it shifts late this year. The gain comes through other people (usually), since it is an eighth house affair.

I looked at your astrological transits. Through the end of 1997, you will be dealing with a transit called Neptune opposite Venus. The ultimate lesson is to raise a shared type of love into a universal type of love. That is, you are called to love for the sake of loving, rather than for the expectation of returned love. Sometimes, people go through periods of disillusionment regarding these issues. This is to help you see clearly the nature of the Universe.

You are also in a time of changes, which call upon you to regenerate and transform in new ways. This will be going on for at least another year. So the easiest way to deal with making changes you already know you want to make is to "Let go and let God." One useful way to allow yourself to move through these changes would be using alternative healing practices, or use of these talents.

Looking ahead to 1998, some of the delays or other things requiring an extra bit of patience could manifest as some sort of reward. The key to this reward is that it comes from things worked for, especially long-term projects. It is a reward for efforts rather than a lottery type of reward. It is a cycle where the Universe wants you to value what comes to you, so it creates situations to help you be clear about who you are, and what you want. The effort brings the rewards. They are long-lasting rewards.

Be well.

Blessings,
Mark Dodich.

My dreams and precognitions seemed to fit Mark's information. His letters confirmed my truth and messages. My lasting rewards were going to come from my new job.
A letter to my sister Emily: June, 1997.

Hi Sis,

I can't believe it has been almost six months since Alex died. Working at the grocery store keeps me busy. It was a blessing to find the job when I needed it so badly. June is the beginning of hurricane season. We try to avoid thinking about it and still

THE GIFT

prepare for it. Fitting shutters to windows, and stocking canned food. Red had said we would be safe this year -- just don't throw away our hurricane shutters.

I need to come home. Danny, when he comes through me in my automatic handwriting, keeps telling me, "When I know the truth."

I don't know what truth. When I come to Ohio, I need to do some research. The energy from his spirit has been amazing. I should have enough money saved by October. I need to find some sort of closure. We all need to be at peace. Jerry, Janet, Henry, and I buried Alex's ashes in the military cemetery near Tampa. I have been working steadily. Princess and I have settled in our apartment. Since I'm alone, she protects me and keeps me company.

I'll call you when I make the travel arrangements.
My spirit guides are working with me, but blind trust and faith are hard. When the time is right and spirit lights my path, I will be there. Peace be with you.

Love and Light,
Carol

SEPTEMBER

The summer seemed to fly by. I had little money for entertainment, since I was trying to save for my trip to Ohio.

I decided to make a trip to go see Red. I hadn't been to The Haven for Spiritual Travelers since the preceding January, when Alex wanted to hear Red's New Year's predictions.

I missed seeing everyone. I borrowed Janet's car for a trip to

◄ **LOVE FROM THE OTHER SIDE**

The Haven. I didn't think to look at the gas gauge. I was on Dixie Highway when the engine stopped. There was a long line of cars behind me. Dear God, why didn't I check for gas? What was I going to do? Halfway down the block I saw a filling-station. Suddenly it felt like an invisible force started to move my car. I was rolling, and steered right into the station and stopped at the pump. *"Wow! Thank you, God, for helping me.* The feeling was similar to an airplane as it lifts off the ground, but the sound was silent. After filling Janet's car with gas, I continued on my path to Red's, drove into the front yard, and parked. I was late and everyone was already inside. I opened the door and walked inside. Red greeted me with a hug.

"We missed you. Is everything all right?"

"Yes. I've been so busy working and I've been through so much. I haven't had the time to come back. It's good to see everyone again."

Red's office assistant, Dawn, looked pleasant in her long, flowing cotton skirt. The purple background with burnt orange and tan flowers had a blouse to match. She had let her hair grow long and her dark eyes sparkled. She greeted me with a warm hug.

When I was a child my family never showed affection, and we rarely hugged each other. When I started attending spiritual and meditation circles, I discovered hugs were shared often. That night, everyone asked me where I had been, why I stayed away for so long.

"Family problems," I answered.

Someone slipped a billet envelope in my hand. I sat down to fill it out. My thoughts were filled with Danny and my upcoming trip to Ohio. I wrote my name for the past, printed it for the present, and my last name would be my future.

I wrote: *I remember Danny's mother. Her tears, how hard she cried, and how much she cared. Has she found peace of mind?*

Red's assistant Dawn opened the service. She told of events and happenings at the home, and a few experiences she had learned from Red's readings.

"Red told me to watch out for troubled waters. Then, while at the beach, I stepped on a jellyfish." She smiled. "It would help if you kept a journal. It is good to look back and confirm a reading Red has given you." She led us in a few songs, while everyone tried to sing in tune. Then Red took over the evening.

"Good evening, folks."

"Good evening, Red."

He spoke of walk-ins and newborn babies.

"A soul can know its parents, but doesn't enter the fetus until it is born," he stated.

I had no need to disagree with Red's speech. But my experience is different. I believe a fetus can feel and know its parents in the womb before it is born. When my mother was pregnant with me, my father worked in the steel mill where he scarffed steel bars — that was the process of turning bars and burning off excess steel and rough spots. The inside of his left ankle was burned when a steel bar rolled over on it. I was born with a scar on the inside of my left ankle. I believe when it comes to spiritual contact or dreams, nothing is written in concrete. The entity or person must relate to their personal feelings and experience for answers.

Red gave the high sign for our ten-minute break. I headed for the backyard to sit on the swing. Before Alex died, I had been coming to Red's home steadily for almost a year. Until now, I had stayed away for nine months. Seeing everyone and being at The Haven felt like coming home. But I knew it would be another long while before I would be able to come back again. Without a car and working weekends, it just wasn't practical. It wasn't long before I heard the cowbell.

Red held my billet in his hand. His hair was tied back. He was letting it grow; he wanted to donate his hair to cancer patients. He looked at me and smiled.

"Hello, Red." I raised my hand.

"Can I come into this vibration?" He lowered his head as if in thought.

"Yes, Red."

"Oh, my God. Yes, this person deserves it and will have it. Bringing a soul like that into the world. She will have it."

"Thank you, Red."

"I have a spirit here. I'm hearing, 'I'm looking over a four-leaf clover.'"

"That's for me, Red. Danny used to sing, 'Roll me over in the clover. Roll me over, lay me down, and do it again.'" I smiled, and so did Red.

"We used to sing that phrase in my Navy days," he said.

Before I left that evening, I stepped into his office to say good night. He was alone and searching through a box of old books. He looked at me. "Yes, Betty?"

"Betty? Red, it's Carol. Why are you seeing Betty?"

"I just want to call you Betty."

Okay, I thought, *I know Red changes vibrations and levels constantly, but he has called me Betty several times. I'm sure Betty is somewhere around me. In time, I'll figure it out. But for now, I'm Carol.* I looked at Red.

"Faithful girlfriend?" he asked, while looking at his books.

"Faithful friend," I said.

"Your path and knowledge are there if you look for them."

He looked tired. We gave each other a hug and I said, "Good night."

Back in my apartment, preparing for bed, I felt Spirit in the corner of my bedroom. I said out loud, "You know, with your negative energy and my positive we could blow this planet right out of the solar system."

The energy immediately left the room.

On the following morning while cleaning and emptying boxes, I unpacked my Hummel statue of the Blessed Virgin and Child. I had picked it out once when Alex and I were in a jewelry store. Alex bought it for me as a wedding gift. It had been packed away since I moved from our home. My apartment was small, and I had no safe place to put it. But as strapped as I was for money, I couldn't bring myself to sell it. For now, I decided to put the statue on my altar beside the rune candle I used for meditation. I didn't know how long I would be staying in my apartment. I had already parted with almost everything I owned.

That night, I received an audio message from the spirit side to give the statue to the priest who had said Mass for Alex. I wasn't sure I wanted to do that. Almost everything I owned had been given away or sold, including my wedding rings. I understood this gift Alex had given me was a part of my past.

Do material things and memories hold us back as we progress toward our future? If the spirit gave me some kind of confirmation, or the time felt right, I would follow through with the action to complete the spirit message.

I prepared my morning meditation. The rune candle with the letter X faced me. I lit a match. As I reached to light the candle, it fell off the altar. I retrieved it, lit it, and took my place on the bed to meditate. While saying my rosary, I noticed the candle was placed on the opposite side and the flame lit up the word "gift." I felt there was a reason for the exchange of energy. My past experience told me to take heed when Spirit spoke to me. I wrote a note to the priest.

Dear Father,

On occasion, when you take the time to appreciate her rare beauty, will you say a prayer for me? Thank you for helping in our time of loss.

Sincerely,
Carol Shimp

I left the statue in the parish office with a secretary. The priest telephoned me the following day.

"Thank you for the beautiful Hummel statue."

"You're welcome, Father. I'm sorry I never was able to meet with you."

"We could arrange it if you would like to talk."

"Thank you, Father, but my work schedule is erratic, and I know you're busy with appointments. I'll pass this time."

"Thank you again, and God bless you."

"God bless you, Father, and thank you for calling. Goodbye."

CHAPTER **25**

Ohio

In October, I called Janet to tell her I had reserved my plane ticket for Ohio.

"Henry is driving me to the airport. Is there anything you want to say or send to your aunt?"

"No, just tell her 'hi' for me. Mom you need a coat. It's going to be cold in Ohio. You can borrow my tan London Fog raincoat. It has a zipper lining for warmth."

"You're right. I do need a coat. It has been so long since I traveled north. Your coat will probably fit me now that I'm skinny. Could you care for Princess? I'll only be gone for the weekend. I could ask your brother, but I would feel better if she was with you."

I couldn't believe I could fit into my daughter's clothes. "If you need anything while I'm gone, just call your brother."

"I'll be fine, Mom, and Princess will be with me. You just have a safe and pleasant trip. Don't worry about the dog."

I pulled a few sweaters off the hanger in my closet, folded and placed them in my suitcase that sat on my bed. Then I found a warm pair of pajamas, and jeans and slacks in my chest of drawers. I felt apprehensive about my trip to visit Danny's grave, wondering what information I might find out or friends I could run into. Would this bring closure to all of my questions? The love energy he sent overwhelmed me; it felt wonderful. But I was so confused. Red had told me that even a snake can be charming, and that I could be hurt. Alex had been sick for so long, in his own bubble of energy. I missed the Alex I had married. Dear God, I tried to do everything right. What happened to me?

At the end of October, Henry drove me to the airport. I decided to wear jeans and a sweatshirt. I knew how cold the air was in the terminal, not to mention transitioning to cold northern weather.

Henry parked in front of my apartment. I opened the door quickly and told him I would be right out. I trusted he would not blow the horn, and wake neighbors at four a.m. He lifted my suitcase into the rear of the truck, and we were off.

"Thank you for driving me to the airport at this early hour."

"It's no trouble, Mom. You've always been good to me. I'm just a phone call away."

It was five a.m. when I checked in at the ticket counter and walked to my departure gate. I sat in the gate area, watching travelers -- children asleep in chairs, and businessmen reading newspapers. I listened to continuous instructions over the intercom. Before long, the announcement was made to board my plane.

It was not a crowded flight, and I was grateful to have two empty seats beside me. I sat by a window and watched the morning sun spread golden rays among the cumulus clouds. My thoughts wandered back -- it had been years since I saw my cousins. I didn't know their children and grandchildren. We double dated and helped each other plan our large weddings. They knew Danny, but I had put a thick wall between us when I married Alex. Here I was, going to visit Danny's grave. The smell of fresh coffee brought me back to reality and I asked the stewardess for orange juice.

Before I knew it, the plane was landing in Atlanta for a short layover. Twenty minutes later I boarded my plane for Ohio. This plane was not full either and left me room to stretch out on the empty seat next to me.

In Ohio, my sister greeted me by the baggage claim area. She had on blue denim slacks with a jacket to match and a white turtleneck blouse. Her hair was cropped short, and I could see she covered her gray with dark brown dye. Her weight was a reminder of how heavy I was before my experience with my Danny's spirit and Alex's death.

"How was your trip?" She smiled, and we hugged each other.

"It was relaxing. I could stretch out on the empty seats beside me. It was a smooth ride all the way." I lifted my suitcase into the trunk of her car.

"Carol, you look younger with short hair. You didn't mention your haircut to me -- I like it."

"Thank you. I had it cut last spring while at the shopping mall on a day off from work. I decided to keep it short."

I thought we could stop at a Chinese buffet on the way home, if you're okay with it. "

"Wonderful. I don't like food at the airport. I only had toast this morning, so lunch sounds really good."

The buffet was excellent with fresh fruit, chicken and rice. I've always enjoyed snow peas. Emily and I refilled our plate for a second helping. We were hungry and quiet while eating our afternoon meal.

"Carol, who drove you to the airport this morning?"

"Henry did. Janet had a long day at the shoe store and Jerry is still getting the kinks out of his truck after his accident last year. I don't think his truck will ever be right again."

After lunch on the way to Emily's house, she filled me in on plans for the week end. I knew she would be driving me around. I was the visitor in her home and she was kind enough to put up with me. I wasn't comfortable bringing my problems home to family. But I needed her.

"Tomorrow, I've made an appointment with a psychic reader

-- in your letter you mentioned reviewing your past and Danny's. So I thought a reader might be able to help give you some answers," Emily said.

I expected it. I knew she belonged to Spiritual Frontiers Fellowship International. She went to readers occasionally, and knew where to find the good ones. With our dreams and sensitivity for energy, it was only natural that we sought communication with like-minded people.

"On Saturday, the family is getting together at our cousin's Patty's home. There isn't time to do a lot, since you are only here for three days."

"I couldn't ask for a lot of time off work. I haven't been there one year yet. The managers were very good to me when they gave me time off for Alex's funeral. I want to call the cemeteries and locate Danny's grave. I need to see it, and try to find completion with the energy that has engulfed my life. I need to do so much in a short amount of time."

"We can do that after the reader tomorrow."

Emily brought me up to date on family events--weddings, birthdays, and who had had surgery. While riding to Emily's home, I enjoyed seeing the landscapes, green rolling hills, and familiar roads that I thought I had forgotten. She parked in her garage; then we carted my suitcase up the steps and into the kitchen. Steps....That's one thing about Florida, there aren't many steps.

Fred stood by the kitchen door. We shared a hug, and then he relieved me of my suitcase and placed it in the extra bedroom. Her large kitchen was as I remembered it. The table at the far wall was covered with notes, letters, and newspapers. In her spare bedroom, I unpacked a few things.

"Sis, where do you keep your phone book? I want to call some of the graveyards and locate Danny's resting place."

"It's in the drawer by the telephone. Help yourself."

"Thanks." I found her telephone book and started to write down numbers of the cemeteries that were listed. On my fourth call, I discovered the cemetery where Danny had been buried. We had a light dinner with ham sandwiches and fruit salad. The evening flew by -- the phone kept ringing, friends and family calling to say hello.

The following morning after our breakfast of toast, grapefruit, and coffee, I put on blue cotton slacks, a navy cotton sweater, and sneakers. I slipped into my daughter's tan London Fog raincoat. It was a gray day with light rain when we left for our psychic reader's appointment. I trusted Emily's choice of psychics. But I was wondering if the information she could give me would be helpful. Emily parked in the driveway on the side of a small house, on a hill. Anne Miller waved from her doorway. She escorted us into her home with smiles and handshakes. Her hands were warm, her energy friendly. Her tiny home was full of books, crystals, and antique furniture. At least, it seemed antique to me. She was pleasingly plump and wore a loose pink-flowered, comfortable-looking housedress. Her hair was short, light blond, and curly.

I picked up a pink pamphlet from a stack I found on the table by the front door. On the front was printed METAPHYSICIAN. On the inside of the first page, I read, "Anne Miller's guidance has been sought by thousands. Her readings have compared with Edgar Cayce. Her comment she had within the pamphlet on this: *My sole purpose is to help others. I know I can because I can see beyond what looks like 'reality.' I am able to give people a new perspective and guide them along their own spiritual path.*" There was a list of credits on her pamphlet: Ordained Minister of Silva, Advanced Reiki Healing, Law Enforcement Task Force, etc.

If Anne works with law enforcement, she must be good, I thought.

READING BY ANNE MILLER 10-26-97:

Mrs. Miller asked who wanted to be read first. Emily said she would be last.

"Emily, make yourself comfortable on the sofa while you're waiting," Anne said.

Emily sat in the front room with books and magazines. Anne escorted me to a small room behind her kitchen. We each pulled a chair out and sat at a round wooden table. Anne handed me a dark blue velvet bag of stones.

"I want you to hold these stones, and when you are ready, spill them on the table in front of us."

I held the blue velvet bag for a moment and thought, *Dear God, please let me find help.* Then, I emptied the bag and discovered polished multicolored stones on the table.

"Do you always use stones for readings?" I asked.

Anne smiled. "Stones have a natural universal energy from within the earth's magnetic field to a higher awareness. I want to choose this red jasper stone for grounding and protection." Anne glanced at the stones and carefully scanned them with her right hand. I noticed a blue agate stone and knew it was for balance. I didn't speak, not wanting to interfere with the vibration of her energy. Then she reached across the table for my hand, and held it.

She began my reading. "I'm feeling a lot of confusion around you right now. It's like I'm going around in circles and I don't know what to do. I feel things need to be taken care of. You might have to have somebody help you. I feel like right on, first things first, second things second, and third things third. You need to get on to other things. You need to get straight on and take care of things the way they are supposed to be taken care of," Anne told me.

"Thank you, Anne. That is exactly how I feel."

"Let me continue." Anne lifted her hand over the stones to feel the energy. "Every once in a while, I get confused and get off the mark. That is when somebody close to you is going to get you back on the mark and get you going again. Things are insurmountable in your life right now. You don't feel like you can get anything together. Nothing is going the way you want it to. This isn't going to last that much longer. I feel strongly first things first and second things second. It's like a filing system. Everything has to be done in order. It's going to be hard for you. You need somebody to help you. Somebody is going to give you a gift that's nice, something that you hadn't expected. What do you do for a living?"

"I work in a grocery store as a cashier," I told her.

"I feel success around it. Possibly a promotion. A little more money. Lots of love around you. I feel for some reason you're leaning on it."

"I feel it is something I have to do."

"That's good. As soon as you get your bearings, you feel like 'As soon as I get on my feet here, I want to tell everybody I'm going to be fine. I just need time.' Things are really going better than your eyes are seeing right now. I feel like I'm going through papers. Things that need to be in order, this is around where you live. I feel like I'm going through drawers and getting organized."

"My journal," I answered. "I keep dates, notes of my dreams and experiences. I'm thinking I might publish them someday. Maybe I can give someone insight from my experience who needs it."

"I want to get it cleaned up. I see you signing papers. There is sadness around it. After the papers are signed, things will start to happen. Doors will open. New friends. Things are good, thank you, God. Is there anything you want to ask me?"

"Yes. I've had visits from the spirit of Danny Malone. I was engaged to Danny in high school. I have had visualizations when he has materialized. I have been zapped over the telephone and his etheric field and energy have been so strong with me. Can you give me some knowledge of what he wants?" I asked.

Anne reached for a red stone and held it, feeling the energy. "What's his full name?" she asked.

"Danny Malone."

LOVE FROM THE OTHER SIDE

"What did you call him?"

"Danny. My husband passed away last January, but Danny came back into my life several years ago. I know my husband is in the light. The light was there in a dream I had, and he comes back to visit me. There is peace there. But Danny, I don't know what he wants from me."

"What was your husband's first name?"

"Alex."

"He is fine. He does check back from time to time to see if everyone is okay. Alex is very busy. With your Danny, I feel doom and gloom over me. There is something he is worried about. I feel him west of where you are. Danny's almost with street people or underground. You won't be able to find him, but he will be able to find you. He will be in touch with you. Bless his heart. Now I don't feel all of the hang-up calls are Danny's. It's almost as if I feel like it was a prankster or something," she said, sitting across the table from me, looking at the crystals spilled in front of her.

"Well, with Danny materializing and walking into people, he is probably using them to make the calls. When I first saw Danny at the flea market -- I thought he was alive, but my sister told me he had a heart attack while playing poker with friends. My guides have told me I'm supposed to help him. But I'm not sure what it is I'm supposed to do."

"Some kind of trouble, I almost feel he saw something he shouldn't have seen. I'm getting wrong time, wrong place."

"Danny keeps telling in my automatic handwriting: *When I know the truth*) I don't know what truth he is talking about. I called the cemetery this morning and asked where his grave was." *Dear God, Anne must think I'm nuts. Does she really know how much I want answers and help?* "I thought if I could feel the energy there, maybe I could bring peace, healing, and closure to his spirit."

"I'm trying to get in touch with somebody in the spirit world that could tell me what happened. Please. There's an uncle. Boy, he's kind of a flippant guy. *He's just saying he's at the wrong place at the wrong time.* Oh gee, you are going to hear from him. Danny just wants to see you. It's like, if he could just be here with you for a short time, that would balance his life. This is like meeting him on the street and he would know where you are going to be. You would think this was by accident. Have you talked to him, or just seen him?"

"I have seen him. I have never talked to him. I communicate with automatic handwriting. I was told in a dream he has the gift," I said.

Anne reached for and held my hand. The warmth of her energy almost brought tears to my eyes. She looked directly into my eyes. "He's got a powerful mind, I'll tell you. You have to keep your protection up, so he doesn't get in and govern you. You could be thinking, well what the heck am I doing this for? He has that kind of control. He is amazing."

"What am I supposed to help Danny with?" I was desperate for an answer.

"All he wants from you is balance. I feel he just wants to sit like we're sitting now. He just wants to be with you. He doesn't look exactly like he did before. It could be a moustache, a different hairstyle or color. There is something different there. You wouldn't recognize him right away until you looked into his eyes. Then you would know who he was."

"You know I can't meet with Danny. I can't find him and I can't get rid of him. How do I get closure to this and get on with my life?" I wanted to find out what I needed to do. My message while I was in the light with spirit was, *You will be protected, he loves you.* How do I ignore God's message? Was it God's light I entered? Where was my confirmation?

"When you finish your prayers and put your protection up for your automatic handwriting, you protect yourself very well, right? Now when you finish, mentally put your mirrors up facing outwards. I have found this to be very effective. Imagine all your mirrors around you. That will take away any negativity and replace it with peace, joy, and love, because mirrors are important."

"Yes, Anne -- I do visualize mirrors." I knew mirrors are used to psychically send an unpleasant psychic attack back to the one who sent it.

"Carol, I just want to mentally talk to him and tell him either to come into your life or stay away -- it's like he has two choices. Danny, come give me information and let me help you, or don't come in anymore, because you're governing Carol's life too much. You need to meet somebody and be happy. You don't need to be brought down by this. And he can bring you down to the lowest depths, because you don't know what to do. Your life is a question. It's really weird because I feel Danny on the other side. But there is a body in that grave. But his mind is just... WOW! It's a fight to keep him out of my head. It's a fight for me because he wants to come into my head and I'm not letting him."

"Well, maybe he has something to say," I said.

"I'll allow him if he just wants to tell you something, but he's not going to govern my life. I won't allow that, and that's where you've got to be strong. Because you've got to. Really, you need to get out and have more fun. Do something different."

"I've met some friends where I work and in my apartment complex. I'm just trying to resolve this situation." I wanted her to know how much I was trying to get my life in order.

"You will do what needs to be done spiritually and psychically when the time comes. I feel protection around you. You have wonderful guides. You need protection, so he doesn't draw you into anything."

"Well, I see him in other people. They come through my line at the grocery store and I get messages."

"Oh, yes. He has that ability. Danny can play with people. He enjoys doing all of this stuff. I feel that sometimes it's the only way he can get through to you, to tell you things. He is amazing. Because he is living on this side more than on the other side, he is between two worlds."

"I have a friend, Red, who has been helping me with this situation in the etheric field."

Anne looked surprised. "Danny is here, teasing, like *Here I am. Ha, ha, ha.* I'm glad you have someone to help you, because if you went to a psychiatrist or therapist with this, they would think you were nuts."

"I know," I said.

"Your friend will help you. I feel him talking to you. He's reaching for your hands, and he's holding your hands. There is going to be a big battle." Anne looked at me with her wide blue eyes. "I feel you need an exorcism, and he will be gone. It's a blessing for him, because he will be where he is supposed to be. After that is done, I feel he will be at peace."

An exorcism? I knew Danny's spirit was inside of me. But, when Anne voiced the word exorcism, I held my breath. I couldn't comprehend the reality of it.

"Then that is what I'm supposed to help him with?"

"I just feel like I'm in the middle. This Red, whom you know, can do the exorcism. I want you to talk to him about it."

"I still need to find his grave. I want to feel the energy there."

"You found out where it is?"

"Yes. I know the section, but I have to go to the office and find out the exact location," I said.

"I don't feel he is going to let you go readily. He is so powerful. That is amazing. You can heal this through your guides and his guides. I want to say, at one point, 'Help me get to his guides.'"

Anne continued, "It's like you would be having a legal battle and we have our attorneys here to fight for us. Take it off your shoulders. You're dealing with something that is up here. It is unseen by us. So have your guides and angels get with his. I just saw a black angel. Just as I said, get with his angels. Son of a gun! That black can be destroyed. Okay?"

"Okay."

"Because I feel your spiritual guide coming over here to enforce and destroy that black one."

"Is there anybody else coming into my life who will have affection for me and wipe all of this away?"

"Not wipe it all away. Because when you meet this other person -- and you will meet him -- then this will already be taken care of. As time goes on here, in less than a year, all this will be wiped away. It could be right now, but it's a pretty big battle. This is a huge power.

"Now you're going to get a kick out of this," Anne said. "I'm seeing this big black cloud over here blowing with big cheeks, and then I'm seeing this white cloud over here blowing. It's like they're blowing each other down. This Red can help you, and he knows somebody else. There's somebody else he knows who could help you, too. You will see Danny face-to-face and it's not that far away, because I can just see you looking at him. When you do, I want you to look him in the eyes and say, "God loves you. God bless you. Goodbye." Don't hang on to it. Because it's hard, the Danny part, the part that was a lovable person. So it's like you would be torn about what you could have loved in that man. But he's not the same Danny you used to love. I almost see his body there in that grave. He's standing beside it laughing, like he's alive, he's here. He could materialize, be sitting here right now, and we would think he was real."

I wondered what gift Danny had received?

"Let me really get into this. How can I see him standing here now? Woo! He's scary. He said, *'You thought you annihilated Christ, didn't you? Here I am.'*

"Woo, boy, that's powerful stuff. But he can be gotten rid of. His spirit is trying to come into me right now. And it's like he wants me to say what he wants me to say. He's not going to do it. I won't allow it. I have to keep telling him no."

"Anne, what does he have to say? Please, I would like to hear it."

"Okay. Just tell me. Stay out of my body. Tell me what you have to say. There again, he's laughing.

"He says, 'You're mine. Nobody else" is going to have you.'

"That's what my guides have told me," I said.

"Okay, now what needs to be done? Here's where we get into first things first. Okay? I would contact Red. You continually talk to your guides mentally to keep Danny away. Oh God, I can't believe this. Even though he's dead, he's not dead."

"Why does he want to hurt me?"

"He's saying right now, 'It's the only way I can get to you.'"

"It's not true."

"Boy, he can be mean.

"He's just saying, 'Don't contradict me. What I say is so.' Well, this is going to stop. This is going to stop now. You're not going to have this in your life anymore. You're going to stop it. Just remember the mirrors facing outwards. He was trying to stop *me* from saying that. Remember the mirrors, and that you are with God every second of every minute of every day. Nothing can come to you but what is for your highest and best good. This has to be removed, and it has to be removed now, so you can get on with your life. And he's laughing now.

"He's saying, 'After the book is finished.'"

We both broke out in laughter. "Boy, this is really something else. Boy, I'll tell you."

"Tell him I owe him at least ONE GOOD SCREW.

We laughed.

"Danny said, 'That's a good one.'"

"This has been a learning experience for me. Because I can usually tune into people, I know if they're passed or if they're here. This Danny completely fooled me in the beginning. That is a new experience for me. I think it's important for you to go to his grave. Also, wear your cross at all times, for protection. And I just want to say some silent goodbyes. Carol, I just want you to look at his grave and say silent goodbyes. And I do mean goodbye -- in all levels of existence."

"Thank you for your help. Just by talking, you have helped

me more than you know. If it wasn't for our spiritual and psychic friends, where would we go for help?"

"It has been a pleasure. What a learning experience for me!"

We stood up and shared a hug. Anne removed the crystals from the table and placed them in her blue bag. Then Emily and I exchanged places and I sat by the magazine rack in the living room.

While sitting in the living room, I thought about Anne's statement: an exorcism. I had been giving Danny grief, trying to send him into the light. He had been with me for several years. The love he enfolded in me was heartfelt. I knew when he left I would miss him. Dear God, why didn't I know what message I was given while I was in the light? I knew my karma with Danny was the fact that I ran out on him. He was hurt. My lesson here: Do not make promises I can't keep.

CHAPTER **26**

Footprints

Outside in my sister's car, we waved goodbye. Anne stood in her doorway and waved back. We drove toward the cemetery. We parked in front of a white building. I stepped inside the office and asked for directions to find Danny's grave. The office assistant checked his computer.

"Yes, its plot number seven eleven." Then he stepped outside and pointed to a curved path. "Just follow the road in a half circle, and you will find it."

We drove around the path and there it was. A huge stone stood above the ground right near the road. I read the inscription from the car window. Danny Malone. 1938-1983. His name was in large letters.

I wondered why his spirit was still earthbound. The year now was 1997. Did he go into the light and then return to earth? How long can a spirit stay on the earth plane? Was there a time limit? How does their level of spiritual existence work? I felt like I was having an anxiety attack, trying to catch my breath as my heart pounded in my chest.

I looked at Emily. "The rain is a light drizzle, but I want to stand by Danny's grave."

"I'll stay in the car." I thought Emily wanted to give me privacy.

I stepped out of the car and looked at the stone. Then I turned around. Behind me, I saw my high school where I graduated. My knees became weak as I remembered the past.

Danny always parked his car in front of the school, waiting for the bell to ring, and for me. I walked out the front doors with classmates and waved goodbye to them as I climbed into the front seat of his car. We would drive to Big Sis for a hamburger and a shake, or to his home or mine. I was taken back forty years in that moment. Tears began to fill my eyes. I stood there in the gray dismal weather. The ground was damp from rain and I was glad Janet had loaned me her raincoat.

At that moment, I thought, standing by Danny's headstone, this might be a Vincent Price scene. Danny chose this gravesite on purpose. Oh God, all of those times we walked from my home to his -- summer, winter, or fall. I remember one time Danny parked his car in some farmer's field. We wanted time alone. We talked and kissed and petted. Then when we wanted to leave, we discovered the tires were stuck in the dirt. He had to call and ask his stepfather to dig us out of the soft ground.

I remembered his stepfather, Joe, looking at us in disbelief. "What in the hell were you doing in that field?" he asked us.

During my high school years, I had a sense of security riding in Danny's car; it was us against the world. But did I go all the way? My father put the fear of God in me. There was no way I was going to become pregnant without being married. I had a fear of displeasing my father, and I didn't want him to lose faith in me. He was six feet tall and two hundred pounds, and always had a deep commanding voice. When he spoke, the ground shook. I found out years later that some of the guys from high school never asked me out because they were afraid of my father.

I did not feel Danny's energy by the grave. It seemed he was watching me from a distance. *Okay*, I thought. *He doesn't like the cemetery.*

I stood by the grave and began to talk to him. "I don't know what you want from me. You can't be with me. You have to go into the light. Someday we will all be together. I need to say goodbye, and I want you to say goodbye to me. We met at the wrong time and wrong place in our lives. I love you and always have. But whatever your problems are, they can be corrected in the light, the source of all power. Please be at peace."

It was time for me to go. Emily sat waiting in her car. Before I left, I noticed a small basket of withered plastic flowers on the side of Danny's grave. A plastic red heart, read "Happy Father's Day." I had never met his wife or children. I could see they cared for him. I wondered if Danny's spirit was also with them, or if they accepted his visitation in Spirit. Silently I read a poem I had written years before.

FOOTPRINTS

How far the distance we did not measure.
Our furring boots in step with our pleasure.
Arm in arm, a friendly warmth we shared.
The breath of our wishes frozen in air.
Snowflakes dressed in delicate lace
Tumbling softly touched our face.
Together we blended in quiet harmony,
Our footprints in the snow held silent memories.

I left my poem there on his grave. I found a small rock close by and placed it on the paper to keep it from blowing away. We left the cemetery. Emily and I drove through our old neighborhood. I remembered Thirty-Third Street and houses where friends had lived. The houses looked the same: two-bedroom, one-bath frame homes. Some homes, friends I knew, had had the attic finished into a third bedroom, and some had a shower in the basement.

I was quite sure all the inhabitants had changed. I was amazed how large the maple trees had grown. The vibrant fall leaves had fallen and left the trees almost bare.

I saw the blue frame home where a neighbor, Mrs. Smith, had introduced me to Alex, and the white frame home where my family and I lived. I had forgotten how large the yard was. I remembered cutting the grass in my bare feet. I could almost smell fresh-cut grass. Then Emily drove slowly up the street past Danny's yellow frame home. I felt Danny's spirit standing in the driveway showing me his home. The energy was intense. Like sitting on top of a Ferris Wheel, and your car starts downward. I remembered when I had called Jane, my next-door neighbor. When we were teenagers, we lay on a blanket in the back yard and tried to soak up the sun. She told me over the telephone that Danny's family had moved. We did not stop by Danny's home, not wanting to disturb whoever was living there.

"Okay," I thought. *His spirit is not ready to go.*

What is it I need to do? Anne Miller had told me that I would do what needed to be done, spiritually and psychically, when the time came.

I didn't share many of my thoughts with Emily. I'm usually a quiet soul. I keep a lot of my thoughts to myself. In reality, if I had discussed my spiritual experiences with any of my friends, they would have thought I was crazy. Fear is what happens to us as we grow into adulthood. It's just the way society has conditioned our minds. I remember what Red told me when a Bible-toting person said they wanted to save him. He said, "I didn't know I was lost."

When they would tell Red, "God does not permit us to talk with spirits," Red would refer to Corinthians 12. *Now concerning spiritual gifts, brethren, I would not have you ignorant. 4 - Now there are diversities of gifts, but the same Spirit. 5 - And there are differences of administrations, but the same Lord.*

We did not stop at Danny's house. I felt pins and needles in my stomach as we drove slowly by. Emily understood. We stopped at the store for a few groceries and drove home. We ate sandwiches for lunch. I found telephone numbers in the phone book, and contacted a few old friends while Emily put a pot roast in the oven. I could hear Fred in the basement, whistling while he worked with his tools.

I spent the next day with the family. Emily and Fred drove me to my cousin Patty's home. When Emily drove into the driveway, I was surprised to see a large tan brick home. Several cars were parked in the circular driveway.

"When did Patty move here? She must have done all right for herself."

"She moved a few years ago. You have been gone for so long; you have a lot of information to catch up on." We entered the house and immediately we received warm greetings.

"Oh my God, I don't even recognize you people. When did you get so old? How long has it been?"

"About twenty years. Time hasn't stood still for you ether. But you're looking good." Patty smiled.

"It's a new generation -- I've been out of touch."

Patty's large dining room table was covered with photo albums. The family gathered around the table. We leafed through photos of vacations, growing up, relatives who had passed away, and memories.

"I have a new granddaughter." Patty said. "And my mother-in-law passed away last year." We ate chips, cookies, and snacks, and drank soda. I caught up on newborn babies, weddings, and idle talk.

"Well, Patty -- you gave me good and bad news all at once. I'm sorry hearing about your mother-in-law. But a new granddaughter, that's wonderful. I don't have any grandchildren."

Time flew by. Before I knew it, I was sitting at the gate waiting for my flight back to Florida. The sky was gray, and the weather chilly. I was glad to return to sunshine, Princess, and work. I gazed out of the window from my seat and my thoughts drifted back to the experience at the cemetery. *Danny,* I thought, *I did not have time to discover your truth.* Maybe I never would.

CHAPTER 27

Back to Work

NOVEMBER

My flight landed in mid-afternoon in southern Florida. I followed passengers through the terminal to the baggage claim area, where I found Janet waiting for me. She had on a blue blouse that matched her eyes, and jeans. Her blond hair was in a ponytail. She smiled and we shared a hug. I was glad to see her.

"So how was your weekend trip? I want to hear all about it."

"It was too short," I said while watching suitcases turning on the conveyor belt. "It had been so long since I saw those people -- they felt like strangers to me. I didn't find the answers I needed. But I did visit Danny's grave. And your Aunt Emily and I went to a psychic."

"So Mom, what happened? What did the psychic tell you? Was Danny's spirit in the grave?"

"Slow down, Janet. I'll tell you in good time. I need to stop and think for a minute." I retrieved my luggage, and we started toward the parking lot.

"Mom, Princess is okay. I took good care of her, and dropped her off at your apartment before I came to the airport. Jerry finished his DUI school and his truck is looking good. He painted it red."

I placed my suitcase in the car trunk. Janet drove toward I-95.

"I'm glad to hear Jerry has finished up his old business. You know, Janet, the psychic, Anne, mentioned an exorcism in my reading. She told me I would be rid of all my Danny problems in less than a year. Later I could feel Danny's spirit watching me while I stood by his grave. He had a large headstone, and it was in a position facing my old high school where he used to pick me up after class. The feeling at that moment took me back forty years."

"Mom, do you think Danny's spirit hurt Dad?"

"I don't know. The thought crossed my mind many times. I can't prove anything. I know your Dad was sick. He had been drinking for a long time. The signs were there. I just didn't want to see them. I was in denial and so was your father."

"So what do you think happened?"

"When your dad was drinking I felt different spirits enter his energy field. Dad left himself open to entities because his guard was down. Maybe Danny's spirit was here to distract me, as a way to help me get through my grief. God knows I needed something to pull me out of my depths of denial."

"Do you think that's why Grandma's spirit was here so often?"

"Well, Grandma gave me the right information in my automatic handwriting. I feel like she was my mediator."

"Wow, Mom. If our path of light and learning is in the stars, I sure hope I find my way. By the way, I'm seeing someone."

"Is it the guy from your Star Trek Klingon ship? The one you mentioned last year."

"Yes, his name is John. I'm so happy. It's a joy to wake up in the mornings."

"Well, I've got Princess. The spirits keep me in suspense. I'm always wondering what surprise I will have next. I brought you and Jerry each a Football Hall Of Fame shirt."

BACK TO WORK

Janet parked her red Camaro in front of my apartment. "Mom, I can't stay long. I know you have to work in the morning, and so do I."

I entered my apartment, set my suitcase on the bed, and greeted Princess. I gave Janet her Hall Of Fame T-shirt, and one to drop off at Jerry's apartment. We said goodbye, and I watched her drive away through the window. Princess pulled her leash off a hook by the kitchen door and was ready for a walk. When we returned to the apartment, I started to unpack my clothes. Knowing I had to work the next morning, I wanted to put everything in order.

I opened the refrigerator and stood in the kitchen with a cold bottle of water in my hand. A sudden force took my arm and threw the water bottle toward the kitchen sink. I was glad the bottle was plastic and didn't break.

Okay, I thought, *I know you're here.* Danny surprised me. The energy from his spirit at that moment had forced me to throw and release my water bottle. I was amazed, but I wasn't afraid. Really, my surprise overtook any fear. I knew he just wanted attention.

I finished unpacking and looked forward to a refreshing shower. When I was relaxed in bed, I retrieved my legal pad and pen. The pen vibrated and my hands were hot.

"Hello?"
"You are the one I want to marry."
"Why did you wait so many years?'
"Because of the way things happened to me. You will know the truth and I love you."
"Do you always give orders? What happened to 'will you'? Instead of 'you will'?"
"You will be my wife. Get used to it."
"You always give orders."
"You know me."
"You have a lot to learn about love."

"I know that you will teach me."

"You're turning my hair gray. Go to the light. I cannot be with you; I want you to be at peace. Why do you insist on torturing me?"

"Go to sleep and dream of me."

I felt a change of energy, like the sun coming out on a cloudy day.

"We all love you in the spirit world and on the earth plane."

"Mom -- are you here, Mom?"

"You will be okay. Mom will help."

"Thank you, Mom, for helping me, and for all of your messages. Things you have told me have come true. Sometimes the timing was off, but I know time in the spirit world is different from our time on earth. Why is Danny so angry?"

"Because he wants to control you. We love you and know that you will be protected."

"Mom, what if I choose not to do this? I mean, who is going to believe this? Our conversations in spirit have been amazing."

"People will believe me. If you choose not to do this, Danny will not give up."

"Thank you for your help and for protecting me."

"Be at peace tonight and rest. Be ready for your surprise."

"I love you. Good night."

It was the first of November 1997, a Tuesday morning. I was dressed in my Publix green and navy uniform. I walked to work to keep my seven a.m. schedule. As I neared the store, I could hear the clanging of grocery carriages as the stock crew rolled them into the breezeway. The tantalizing aroma of fresh-baked bread drifted from the bakery. I entered the store, and waved and nodded a good morning to other associates. I found my way to my assigned register. I asked for my cash bag and counted the money in my till.

"Hi, Joe." I greeted one of the older men who packaged food orders on my register.

"Good morning, Carol. I've been waiting for you to come back from your trip."

He handed me a piece of paper. Joe was a pleasant gentleman. He was maybe five feet tall, chubby, with white hair and a receding hairline. He wore the company green T-shirt and navy slacks. I looked at the strip of paper -- a telephone number. Joe smiled.

"There is a regular customer who wanted to get in touch with you. When he asked me about you I told him I knew you were a widow. He wants you to call him. He's a nice man; I've been helping him pack his groceries for years. I know you'll like him. Give him a call."

"Thank you, Joe, for thinking of me."

Joe triggered my curiosity. I looked at the numbers. *Nice writing*, I thought. Alex had passed away last January. It was now early November. I missed companionship and I felt this emptiness inside. I thought with the holidays coming it would be nice to have a friend. I put the paper in my pocket. I wondered what he looked like and how many times he had walked through my line.

"What does he look like, Joe?"

"Well, he has white hair, a mustache, and a very nice smile."

The lights became bright and I heard the sound of sliding front doors as customers straggled in. I watched the group of elderly friends from the Sunshine Home for the Retired. Most of them looked to be ninety years old. I just knew that on this morning I was going to be tested by the Lord. Tonight I would write this experience in my journal. *All of my little notes*, I thought. *Will I ever have them in order?*

It wasn't long before they began a line at my register. Etta, a frail elderly lady wearing a pink straw hat with flowers on it, started placing items slowly, one at a time, on my conveyor belt. She had a purple floral blouse on with purple slacks to match. I scanned her items -- bananas, prunes, a bag of hard candy, a small package of hamburger, bread, and a few other things, while Joe packaged her groceries.

"Young lady, don't you bruise my bananas," she commanded.

"I was very careful with your bananas. That will be fifty-nine dollars and twenty cents, Etta."

"Oh my goodness, I didn't realize I spent so much. Your prices are going up and up." She rummaged through her worn black cloth purse. "I can't seem to find my card." Etta stood there with a confused look on her face.

Joe looked at me and gave me the smile I anticipated.

"Take your time, Etta. You'll find it." I tried to calm her frustration.

"Oh, here it is. I'm sorry to hold the line up. You people are so nice and patient with old folks like us." Etta pulled a card out of her purse.

"You're fine. Just slide your card and you can be on your way."

"Okay." Etta slid her card.

"Etta, your card's not working. It must be a credit card."

"Oh no. I don't have any credit cards. Let me slide it again."

"Etta, it's still not working. Do you have a pin number?"

"No. I have never had to use a pin number. Let me slide my card again."

"Etta, if you don't have a pin number then it has to be a credit card."

"You're making me all nervous, I don't have a pin number and I don't use credit." She was about to become really upset. I noticed a tear in her eye.

"It's all right, Etta -- I'm going to store this order in my computer and Joe will walk you up to customer service. The manager will take care of you." Joe held Etta's arm and wheeled her grocery cart toward the office.

"Well, now I suppose I'll have to pack my own groceries!" a short, thin elderly gentleman with gray hair complained loudly. He waved a hand in front of me and shoved his groceries on the conveyor belt toward my scanner.

"I'm sorry you were held up, sir. I'll pack your food for you."

BACK TO WORK

Just then, a manager walked over to my work area. Eric was thin and always neatly dressed. He had short wavy red hair. It was not the real bright carrot red, but a slightly darker shade.

"Is there a problem here? Can I help?" Eric smiled.

"Well, this nice gentleman needs help with his groceries," I explained.

"You people need a better system. My milk is turning sour," the customer told Eric.

"Let me help you, sir." He started to bag the groceries. "I'll walk you out to your bus."

Eric was always helpful. His timing was excellent. Andy the elderly white-haired gentleman was next in line. I had seen him many times. Andy was always smiling and in a good mood. For an older person, Andy always dressed neatly. He had on a new yellow sport shirt and tan slacks.

"How are you doing this morning, Andy?"

"Well, I'm on the right side of the grass. I had to read the obituary column this morning to see if I should get out of bed." He smiled.

"Andy, you are always so cheerful. The rest of my day will be a happy one."

"Carol, I wouldn't give my business to anyone else when there's a pretty girl like you working."

"Why thank you, Andy. It is a pleasure waiting on a handsome young guy like you."

Most of my elderly customers were pleasant. As the day continued, it became more agreeable.

CHAPTER **28**

Thomas O'Leary

On my next scheduled day off, I walked to the neighborhood Laundromat, cleaned my apartment, and bathed Princess. There was a hose attached to the outside of my apartment building. Princess liked her bath and did not fuss. She shook her fur and sprayed soap all over me. When I looked in her eyes I swear she was smiling. I showered and ate a snack.

Then, I gathered my courage and picked up the phone. On the fourth ring, I replaced the receiver. He must not be home. Princess and I went for our evening walk before bedtime.

On the following day, I was back at work.

"Good morning, Joe."

"Hi, Carol. My friend, Tom, tried to return your call yesterday evening. He told me your phone number was listed on his caller ID."

"How did you know I tried to call him, Joe?"

"He stopped in yesterday evening for a few grocery items and told me."

"I must have been walking my dog, Joe. What's his name?"

"Thomas. Thomas O' Leary."

Oh boy, another Irishman.

Two days later, I tried calling him again, but there was no answer. *Well*, I thought, *he knows where I work. He can come see me. Is he shy,* I wondered, *or is this his game?*

The following evening, I was relaxing at home writing in my journal, when I was startled by the ring of my telephone.

"Hello?"

"Hey, I finally made a connection. It's Thomas."

"I tried to call you. You must have been out, Thomas. I was surprised when Joe handed me your phone number."

"Joe gave you a good reference. I'm sorry if I made you uncomfortable while I watched you work."

"That shows how much I pay attention, I didn't notice you." The sound of his voice was pleasant. I wondered what he looked like.

"I would like to take you out. Is there anywhere special you want to go?"

"You know, the county fair is on. I haven't been to a fair in years."

"That sounds like a date, Carol. I'll pick you up this Saturday night at seven. Where do you live?"

"It's a white two-story building behind Publix. I'm in the second apartment from the east side." I gave Thomas my address, said goodbye, and replaced the receiver. I hadn't been on a date in almost forty years. I almost felt like a teenager again. I stepped into my small walk-in closet and scanned my wardrobe, trying to decide on a pair of jeans and cotton shirt. It was Monday. I had nearly a week to think about my big night out.

Would Thomas have come through my checkout line? Thomas, Tom, Tommy? I wondered what name he used.

Dawn, Red's female assistant, came through my line that week.

"Hi, Carol. We miss you on Sunday evenings. Red thinks of you often. Where are you living now, and how are you doing?" She smiled and her dark eyes sparkled.

"I have a little apartment not far from the store. My dog, Princess, and I are comfortable there. Tell Red for me that Danny's spirit still hasn't gone to the light. My work schedule is erratic. I haven't been able to visit. When I have a break from work, I'll surprise you one Sunday night."

"We look forward to it. I just stopped in for a few items. I hope we can get together soon." Dawn picked up her bags and left.

On the third morning of that week, I lay in bed, reluctant to wake up from a sound sleep. I began to open my eyes and drew in a deep breath. When I exhaled, a spontaneous gray smoke, like a vapor, drifted out of my mouth. I readily believed from my experience with Red, that he had helped me from afar. Now that I thought about my experience, I wanted more information. I wanted to share my incident with Mark and discover his feelings about it. I wrote to Mark because I wanted to be able to read his response. The following morning I mailed my letter on the way to work.

I received a response the following week.

Hello Carol,

You asked me to respond to your experience, releasing the vital life force. There are a variety of sources you can try to find information. Start by searching the words "prana" and "manna," as they are life force from a Hindu perspective.

I have seen accidents that release life force and also accidents that bring in a greater aspect of the soul. I don't really know the reason why it happens. Although I would take this information lightly:

Louise Hay says that accidents have to do with an inability to speak up for oneself. And since the release was through your mouth, perhaps there is something to it, especially since you started writing as a way to reclaim your voice.

> *I do know that shamanic healers do a process called "soul retrieval" to help people pull back that part of them that was scattered in the universe. Nothing is really ever lost. If you look into this path, be sure anyone you work with is walking a path of heart and integrity. Shamans tend to be very kind and loving people. Or people caught up in power and control. I hope this helped to answer your question.*
>
> *Be well, Carol. Blessings.*

Mark's letter helped to clear my mind, and the stress I had been feeling about my experience. I read his letter again and the word "shaman" stared back at me. I knew Red was a shaman. In my research I found out, when a shaman locates misplaced energy, he or she will then merge with their guardian spirits or power animals. This increase of power allows the shaman to pull the misplaced energy out, as if he or she has a strong magnet to pull with. If the shaman's energy is powerful enough he can pull out the displaced energy's hold on the body. Red had been giving healings for the thirty years that I had known him. He wasn't even in the room with me. I didn't know someone could remove energy by proxy. This was no experience like Linda Blair's in *The Exorcist*.

From his home, I believed Red had released me from displaced energy. I couldn't believe that I was finally free.

I sent a thought. *"Thank you, Red."* I realized when Dawn, Red's office assistant came through my line in Publix, Red found the primordial force and set into motion the particle of the atom by mental – emotional activity. Mentally and by proxy, Red was able to help me.

I had experienced a release of the vital life force. The Breath of God. All I could think of was *thank you, thank you, thank you*. My experience with Danny started at the flea market on Mother's Day, after Spirit lifted me into the light on Easter morning.

I had hoped my release of displaced energy would help Danny toward the light. Later I found out that Danny's spiritual energy was still in my apartment and I felt his energy in my bedroom. I don't know how to describe the feel of spirit energy. But like James Van Praagh, or Mary Ann Winkowski, I knew it was there. I remembered Anne Miller in my reading told me it might take more than one episode to finally release Danny's spirit.

While working that week I kept looking for Tom. Several men with white hair came through my checkout line. But I had not caught sight of Tom or anyone that I thought might look like a Thomas in the store all week.

On Saturday evening, I decided on a red cotton blouse and a pair of blue jeans. At seven p.m., Princess had her nose in the front window between the vertical blinds. She yelped a greeting to let me know we had company. Thomas O' Leary was knocking on my front door. I opened the door and found a blue jeans and white T-shirt kind of guy. His short hair and moustache were white. He had a pleasant smile and the kindest blue eyes. I just reached his hairline. He was my height -- five feet four inches, with a stocky build. When I looked in his eyes, I had the feeling I had known him before. In reality, I knew it was my first meeting with Tom.

"Hello," Tom said. "You look terrific."

His compliment made me smile. *I looked terrific, wow.* He reached for my hand. "I'm sorry we have to use my work truck. It's the only transportation I own."

Princess was immediately by the door to give her approval. Tom petted her enthusiastically, and they became friends immediately.

"That's fine. I don't mind at all." I locked the front door and stepped toward his white 1994 Dodge pickup truck. Tom opened the door for me and gave me his hand to help me up the step. I felt the rough calluses on his knuckles. Somehow, I knew, subconsciously, that I had been waiting for him. He drove down the road toward the racetrack and the site of the county fair. We talked.

"So tell me about yourself, Tom. Where do you work?"

"I work with marble, and at times, ceramic tile."

"Really? My husband used to lay tile on pool decks and driveways. Do you have any children?"

"I have a daughter, Marie, and a son, Jerry, both born in June."

"Both of my children were born in June." Wow, I thought, *there seem to be a lot of similarities. I feel so comfortable in his company.* "Are you divorced or widowed?"

"I'm divorced. It's been almost eleven years now. It was pretty nasty."

"I lost my husband last January. He was hit by a truck on the highway." I didn't mention my husband often. Not even to co-workers -- the memory was too painful. My throat tightened and I choked back a sob.

"That's tough."

Tom told me he lived on the east side across the road from the store where I worked. Tom said, "I'm Catholic, but I only attend Mass on Christmas and Easter. And I don't drink or smoke."

"Tom, I'm Catholic, but without a car and working weekends I haven't been to church in almost a year. Are you a native of Florida? Just about everyone here is from somewhere else."

"I'm from Pennsylvania, near Philadelphia."

"Really? My mother grew up in Pennsylvania. I grew up in Ohio, near Cleveland."

The parking attendant waved his yellow flashlight, directing Tom into a parking place. We walked toward the fair and Tom purchased tickets at the entrance. He held my hand, and we walked through the gates. I smelled cotton candy and grease from the funnel cakes. We heard calliope music, and the children were lined up by the merry-go-round. There were sounds of laughter and screams of excitement as we passed the bumper cars.

Tom continually discouraged vendors by waving his left arm, and then nodded his head in greeting.

We passed a food stand. "Gee, that fried chicken smells good," I said.

"Oh, Carol, I don't care for chicken. Do you want some?"

"No, that's okay. We'll find something else."

Wow, I thought, *Alex didn't care for chicken, he worked with tile, and our children were born in June. Tom was from Pennsylvania, he was Catholic, and he didn't smoke or drink. In my automatic handwriting, Mother told me I had a surprise coming. I wonder if she meant Thomas?* We decided on hot dogs.

We walked toward the bright colored lights on the Ferris Wheel.

"Would you like to ride on the Ferris Wheel? I love to sit on the top and look over the events of the fair."

He looked up, toward the lights and top. "No. Carol, I'm not fond of the Ferris Wheel."

Now that was just too many coincidences. Alex was afraid of heights. In the livestock barn, the smell of sawdust and livestock engulfed us. We walked through sawdust. We petted the rabbits in their open cages and checked out the pigs in their pens, chickens, and other animals. I enjoyed being with Tom. His energy felt good. I was comfortable with him. I discovered Tom was two years younger than I.

On the following weekend we saw the movie *Titanic.* Tom purchased hot buttered popcorn and a soda for each of us.

A few weeks later, my daughter invited Tom and me for Thanksgiving dinner in her apartment. Janet's dining area was larger than mine. Janet and her new squeeze, John, cooked the whole feast. John peeled potatoes and helped with the turkey. Janet even made pumpkin pies. Grandma's china was placed on the table. Jerry tormented his sister, as usual.

"I can't wait to taste the pies my twisted sister made." He stood beside Janet, inhaling the aroma of fragrances in the kitchen.

"Don't hold your breath, brother. You don't look good in blue."

LOVE FROM THE OTHER SIDE

The television showed burning fire logs on the screen. The aroma of cinnamon candles set the scene for friendship and spiritual warmth.

Three weeks had passed, and I gave Tom a key to my apartment. A feeling of warmth overwhelmed me when I was with Tom. I had built confidence and trust in him. He offered to walk and feed Princess when I had to work late. It was a relief for me not to have to rush home and take care of Princess. I felt someone could know a person for years and be sorry they trusted them. But my feeling was that he was honest. I would come home from work and there would be a note on my dining room table. *"I fed and walked the dog. Love, Tom."*

Another evening, after working all day, I came home and I found my dirty laundry all washed and folded on my bed. I felt embarrassed, because Tom had laundered my personal undergarments. We had become close, but I didn't feel that intimate toward him yet. Time was moving so fast. Tom had lifted my burden of walking to the nearby Laundromat. I sighed. Under my breath I said, *"Thank you."*

He invited me to his home to meet his family. His house was on a white concrete foundation, with a front porch and two bedrooms. The large living room had a corner fireplace. His daughter Marie was there eating pizza with her son. Marie was slim and had blond hair. Her son looked to be about nine years old. Tom's son shared my son's Jerry's name. He was well-built. I was told he went to the gym a lot. He had light-brown hair and a buzz haircut.

"Tom, it could become confusing – your Jerry and my Jerry," I said.

"No problem. We'll work it out." Tom put his arm around my shoulder.

Then Tom drove me to his brother and sister-in-law's home. I was surprised. His brother lived only a few blocks from Tom's house. We could have walked. We stood outside the front door and Tom knocked on the doorframe.

"Come in." We heard a tenor voice.

We stepped inside, and Tom introduced me.

"Carol, I want you to meet my brother, Paul, and his wife, June."

They rose from their chairs and greeted me with handshakes and hugs. As I walked into their living room, I had the feeling I had been there many times. Was it maybe in my dreams, or was it an out-of-body, astral projection?

The carpet was light pink. Placement of a tan sofa and nature paintings on the wall left me with a pleasant sensation. The living room and dining area were open. Paul was around six feet tall, with white hair like Tom's, and a nice smile. I felt I had known his sister-in-law, June, all my life, although this was the first time I had met her. Some people just instantly feel like family. June was thin, maybe five feet seven inches. She had short brown hair.

"Can I put on a pot of fresh coffee?" June asked.

"No, thank you," Tom answered.

"Well, I didn't ask you. I asked Carol," June said.

"It's all right, June. I'm going to cook dinner for Tom after we leave."

We sat for a short while. Tom bantered with Paul about NASCAR drivers and races.

June and I discussed makeup and perfume, and leafed through catalogs. Then we said our goodbyes.

That evening, back in my apartment, I started to clear paper and notebooks from my glass dinette table. I decided to set my table with the few plates I had left from my life with Alex. My blue placemats looked nice with blue and white plates. For a final touch I lit a tall white candle in the center. I gave a last glance to be sure everything was perfect, and I went back to fixing a steak dinner with baked potatoes and a salad.

"What was all the paper on your table?" Tom inquired.

"It's my journal. I'm writing about spiritual experiences and dreams. Someday I might decide to publish them."

"I hope they're all good ones."

I fed Princess so she wouldn't bother us while we ate. After an enjoyable dinner we had an evening stroll with Princess. The sunset was beautiful with reds and golds in the sky. Once we returned to my apartment Tom helped me with dishes. Not long after clearing the table I stepped into the bedroom to put my slippers on and spotted Tom's work boots planted beside my bed. He must have taken them off while I washed the dishes. We had been dating for almost two months. I knew Tom wanted more commitment in our relationship. Alex was my first and only lover. I had never been with anyone else. I cared for Tom and knew I had to make an effort to keep him. I had a nagging feeling I was not ready for sex with anyone. But I knew I wanted to try with Tom.

He entered the bedroom and enfolded me in his arms. We kissed -- long, slow, and tender.

"Let's take a warm shower together," I suggested, hoping it would excite my sexuality. *Was I ready for this?*

"What a wonderful idea." Tom lifted my blue cotton shirt as I raised my arms. He pulled it clear of my head and laid it to one side before starting to undress himself.

We stepped into the shower under the warm water and gently and carefully began to wash each other. We laughed, feeling our movements stir and awaken our senses. Timeless moments passed before we stepped out and I handed Tom a towel. Quickly we dried off and fell into bed. I was grateful I hadn't felt Danny's energy. His spirit must be busy doing whatever spirits do. *Thank you, God – he is not here with Tom and me.*

Sex was nice. Tom said it was great. But for me it was not complete. I was used to years of Alex's energy. That first wonderful feeling I had with Alex I knew would never be the same with anyone else. When you have sex with one person for so many years, it's hard to open up completely, especially the first time. Tom though, was caring. I knew I would grow to love him more and more.

CHAPTER **29**

Christmas

I came home from work on Christmas Eve, and what a surprise! I found a tree in my dining area, decorated with multi-colored lights, red bows, glass bulbs, and gold garlands. A white teddy bear sitting on a sleigh decorated my coffee table. I couldn't believe Tom did this for me. I wanted to call him, but it was late and he was probably sleeping since he had worked most of the day.

I had Christmas dinner at Tom's house. He used paper plates. He cooked creamed chipped beef with white rice and California blend vegetables. It was a stir-fry concoction, and it was good. My children had dinner with their friends. It was the first Christmas dinner our family hadn't celebrated together. Later that evening, our families all gathered in Tom's living room. "Have a Holly Jolly Christmas," was playing on Tom's stereo. We had eggnog, cookies, and hot chocolate.

"Your Christmas tree is beautiful. Did you decorate it?"

It held white doves, cream-colored velvet ribbon, antique bulbs, and white lights. The Christmas tree stood by the fireplace. It made a perfect greeting card scene.

"Yes. My mother would have liked it. Most of these ornaments belonged to her. She passed away a few years ago. She loved Christmas."

"Well, you also did a beautiful job decorating the tree you put in my apartment. You're talented."

"It was a pleasure to trim the tree for you. I want you to be happy."

"Thank you, Tom. With the grief I've been through, I almost forgot happy. What was your mother's name?"

"Betty."

Well, I thought. *I'll have to tell Red that I finally met Betty.*

We were becoming close. On New Year's Eve, Tom drove me to the beach to watch the sunset's reflection on the ocean. We found a bench under the palm trees and I watched a blanket of crystal confetti reflect on the water. I removed my shoes, and my bare feet imprinted on the warm white sand. As I walked toward the ocean's edge I avoided stepping on a box crab digging his tunnel in the sand. Tom followed me. He wrapped his arms around me, looked in my eyes, and asked, "Will you move in with me?" A flock of seagulls flew by and their squalls sounded approval of the day's end.

"I would love to, Tom. But I don't want to marry again. Not now and maybe never."

Is there an off switch to lessen pain in my heart?

With my right hand, my fingers lightly brushed the empty space on my left ring finger. I was afraid, because my possessive spirit was still haunting me. I did not want to go through with Tom what I had experienced with Alex. Tom, I knew, didn't drink and Alex did. Maybe if I found happiness, with Tom, Danny would move into the light. *Was that what he wanted?*

"Don't worry, Carol. We'll take one day at a time." Tom held me close and kissed me.

I felt warm and protected in Tom's arms. Under my breath I prayed that happiness would follow. Would Tom understand my spiritual experiences and the other world I was connected to?

Would Danny be in my way? Could I finally find peace? After Tom left me on my doorstep I had so many apprehensions about my future. I believed I needed a regression.

I had found a friend who referred me to a Mr. Bob Decker. I knew Red would have given me a regression. Red had helped me tremendously, but I felt Red knew me too well. I thought new energy would be the key. Also, I felt the more people I had to establish my situation, the more believable my story would be.

CHAPTER **30**

Regression

Hypnoses, Regression For Carol Shimp, By Bob Decker – 28 Febuary 1998

I knocked on the front door of Mr. Decker's townhouse. I wasn't sure what to expect. I had been to readers and psychics. They were all helpful and correct with their information, but I still had my troubled spirit with me. I had hoped against hope that this time I could find closure. Bob Decker answered the door. He must have been around five foot five. He might have been in his late forties, and was just a little pleasingly plump around the waist. His hair was short, light brown or maybe dark blond. Mr. Decker invited me into his living room, motioning toward a large black recliner. I sat down and positioned myself comfortably. Mr. Decker placed a tape recorder on the table near me and immediately programmed his voice to a monotone vibration.

"What are you looking for in this regression, Carol?" he asked.

"I need to find peace with an earthbound spirit that I've had experience with in this life."

I immediately began to fall into an altered state.

Recognizing this, Bob said, "Tell me what you perceive, Carol."

"Light. I see light."

"Very good. You're doing very well, Carol. It's time to call out to your highest guides. Call out for them; call out to the ends of the Universe. Call out for them and know that they will come to you, If they haven't already."

Time passed. ------I mentally tried to call my highest guides.

"And at this point, Carol, your guides are there. They are with you right now.

It's time to find out their names. I want you to trust your impression. You're going to get their names one letter at a time. I'm going to go ahead and count from one to three. The first letter of their names will come into your mind at three. Just trust me. One, two, three. The first letter is?"

"J."

"And the second letter is?"

"O."

"And the third letter is?"

"B."

"And the fourth letter is?"

"I don't know."

"That's all right. You're doing very, very well. His name is Job. You have his name. Go ahead now, Carol. You have his name. Repeat the name over and over in your mind. Hear it echo out to the ends of the Universe."

Time passes. ----------- I'm trying mentally to call out his name.

"And at this point, Job is appearing before you. Job is appearing before you. I'll count to three, and you will be able to perceive him clearly. One, two, three. Now that you clearly perceive Job, tell me what you perceive."

"My book."

"You see your book?"

"Yes."

"What color is the book?"

"White."

"How big is it?"

"It's like a dictionary."

"All right. The book opens up now. The book opens up. Can you tell me what you perceive in writing? Can you make out any of the writing, Carol?"

"My will is done."

"Very well. You're doing very, very well, Carol. Now let's go ahead and call upon your masters and teachers. Go ahead and call upon them by name, if you know their names; the particular masters you are attached to. Go ahead and call them now."

Time passes. --------I'm mentally trying to call my masters and teachers.

"And now tell me what you perceive, Carol?"

"There are many."

"Good, very good. How do you feel in their presence?"

"Balanced."

"Very good, good, very good. Enjoy that sense of balance. Drink in their energy. Partake of their energy. Let that balance fill you on every level -- mind, body, and spirit, Carol. So you have your spiritual masters around you -- your guides. You have the support group to assist you. At this point, if there are any deceased relatives or loved ones with whom you are comfortable -- having them with you -- you may call them at this time."

"No."

"Okay, okay. Very good. Very good. At this point, you have assembled around you a spiritual support group. And this support group is going to empower you, balance you, and protect you. As you explore your past, present, and perhaps -- if necessary -- your future. As well as your Karma, your Dharma, and all that motivates or blocks you in the present. For now, I'm going to go ahead and ask you questions. And you are going to repeat what they say.

"They will speak to you in thought language. I want you to just simply trust your impressions. You'll repeat it back to me, so we can record the response. I ask the spiritual support group: What is the nature of the relationship between Carol Shimp and Danny Malone?"

"Love."

"All right, there's love. But who loves whom more?"

"He. He loves me."

"Danny loves more, doesn't he? And that's okay. Yes, we understand that. And I ask the spiritual support group, does Carol have a past life connection with Danny?"

"Yes."

"We understand that, and we accept that."

"Yes."

"Okay, I'm going to count from one to three, and at three, your spiritual support group -- the spiritual team -- is going to protect you, and take you back to a past life, an important one for you to see when you were with Danny. Trust your impressions. It's all right. You're safe and protected, Carol. One, two, three. Go there now. Tell me what you perceive."

Time: *I'm trying to go back through time.*

"Are you indoors or outdoors?"

"Outside."

"Okay. Describe the scenery to me. What does it look like out there, Carol?"

"Like a mountain."

"Like a mountain, you say?"

"Mountain."

"Okay, okay. And is there anybody with you at all?"

"There's a child."

"Is it your child? Does the child belong to you?"

"Yes."

"Is the child male or female?"

"Male."

"Okay. How old is he?"

"Four."

"Okay. So you're in the mountain region with this four-year-old. Where's Dad? Where is his father?"

"He is with me."

"He is with you, too?"

"Yes."

"Okay. All right, all right. I'm going to count from one to three, and at three, vivid impressions will flood in with the name of the place that you are now, at least the country. Trust your impressions. One, two, three: where are you now?"

"The mountain."

"Okay. Very good, very good. I'm going to count from one to three and vivid impressions will flood in of your name, in this past life that you are now experiencing. Trust your impressions. You have to really trust. One, two, three. What's your name?"

"Myra."

"Okay, Myra. And what's the little boy's name? I'm going to count from one to three. Trust your impressions. Just trust your impressions. The information will flood in for you."

"Jacob."

"Okay, okay. Myra and little Jacob. And what's your husband's name, Myra? I'm going to count from one to three. Vivid impressions will flood in, one, two, three. What's your husband's name?"

"Harold."

"Okay -- so Myra, Harold, and little Jacob are in this mountainous region. Let's go ahead a little bit and explore this together. I'm going to count from one to three. At three, vivid impressions are going to flood in of your daily life together, in this past life you're now experiencing. This could be just a routine scene inside the home. Your daily routine -- vivid impressions -- will flood in for you, Myra. One, two, three. Tell me what you perceive."

"Different colors of cloth."

"Okay. Do you work with cloth for a living?"

"Yes."

"And what does your husband do for a living? What's his occupation?"

"He's a healer."

"Ah, healer. Wonderful, wonderful. And Myra, as you're examining this life, do you see anybody whom you recognize from your present incarnation as Carol?"

"No."

"Okay. Let's go ahead a little further. Let's go ahead a little further in this life. Let's go ahead and see what happens to Jacob, your son, as he grows older. Let's go ahead a little further. Now, at this point, the boy is growing up. The boy is growing up. I'll count from one to three. Vivid impressions will flood in about Jacob. Tell me what you perceive."

"He's a hunter."

"Ah, he's a hunter. Is he good at it, Myra?"

"Yes."

"What does he bring home?"

"Rabbit."

"Rabbit. Oh, that sounds tasty. Anything else?"

"No."

"Now, you said your husband was a healer. I'm going to count one to three. Vivid impressions will flood in of his healing work, what you witnessed of it. One, two, three. Vivid impressions flood in. Tell me about his work as a healer."

"Colors, energy, heart."

"Okay. So he works a lot with the heart energy, colors. Is that what you're saying, Myra?"

"Yes."

"Is he very popular as a healer?"

"Yes."

"How do you feel about this? There must be many people coming to him. How do you feel?"

"That's why we're on the mountain."

"Okay. So, it's not like you're having a huge amount of people coming. It's like local villagers and other people who live around the mountain with you."

"Yes."

"And by the way, I'm interested in what year this is. I'm going to count to three. At three, I want you to really trust your impressions. One, two, three. What year is it?"

"1670."

"1670. That's fine. Is it B.C. or A.D?"

"A.D."

"A.D. All right, so here you are in this mountain region. The boy's a hunter; your husband's a healer. Let's go ahead. Let's go ahead, and go on up to the last day of this life that you're in now, Myra. You're not going to experience any pain or suffering, because you're only going to watch it as a movie. That's right. It's a command. You're only going to watch it as a movie. You're going to obey. You're just going to watch it as a movie, without pain or suffering. You haven't crossed over. I'm going to count from one to three. You're going to the last day of this life. You're going to watch it as a movie. One, two, three. Go there now, and tell me what you perceive."

"Bows and arrows."

"Okay. Who has the bows and arrows?"

"Men."

"Did they shoot you with bows and arrows?"

"Yes."

"Are Jacob or Harold there? Did they get shot, too?"

"Yes."

"So they die as well."

"Yes."

"Okay. All right. Now, I'm going to count from one to three. You will have crossed over. You will have crossed over. One, two, three. You have crossed over. Where are you in relationship to your body?"

"I'm in a void."

"Okay. Do you see your body? Are you above it, or off to the side of it?"

"Off to the side of it."

"But feel the peace of the spirit. You're free of the flesh. You're free of that body full of wounds, bows and arrows. You're free. Feel the freedom, peace in spirit. There's no suffering here. There is no suffering here, only peace. Take a moment, and feel that. Take, a moment and feel that."

Time passes. ------

"All right. I'm going to go ahead and count from one to three. At three, you're going to come back to the present. Keep your eyes closed. Stay relaxed. You're going to be back with the spiritual support group. You're back with the spiritual support group. Job will be there. Guides and masters will be there. All surround you with love. One, two, three. Are you back with the spiritual support group now, Carol?"

"Yes."

"Now I'm going to ask the spiritual support group, was Danny Malone in that past life with her when she was Myra?"

"Yes."

"Okay. Okay. Which person was Danny Malone in that past life?"

"He was my husband."

"Okay. Okay. So what was the nature of his relationship with Carol that carried over into this life?"

"Loss of love."

"All right. So that painful separation, that painful separation, that painful parting in that past life, has carried over into this life. Is that correct?"

"Yes."

"Okay. I'm also going to ask Job and ask the spiritual support group: Are there any other past lives where there was an abrupt and painful parting between Danny Malone and Carol Shimp? Trust your impressions, Carol. Tell me what they say, or what they show you.

"Tension."

"Spiritual support team, can you explain that a little bit more please? What's the nature of the tension? What is the true nature of this situation?"

"I can't go there."

"Why is that? Why can't you go there?"

"Pain."

"Okay. If you could go there without pain, could you do it? Of course you can. Spiritual support team, show it to her as a movie. Bring up a screen right now. Do you see the screen, Carol? Do you see a movie screen in front of you here on the platform?"

"Yes."

"Let them show it to you as a movie. Just only as a movie. I'm going to count from one to three. Tell me what you perceive. It's only a movie screen."

"Negativity."

"Okay. What kind of negativity? What's happening up there?"

"Witchcraft."

"Who's doing the witchcraft?"

"Danny."

"Danny Malone, you say?"

"Yes."

"Are you involved in the witchcraft?"

"I can't go there."

"All right. All right, let go of this. Let go of this, let go of this. The movie screen goes blank. You don't have to deal with the movie screen anymore. You don't have to deal with this. It just simply goes away. Is it gone?"

"Yes."

"Okay. Okay. Evidently, in a past life, you had an abrupt separation with Danny Malone. You had at least one past life where you engaged in negative magic along with him. Now you can't take that back. What you did -- what you did -- is gone and it's done. But how you carry it into the present -- whether or not you forgive yourself for it...that, my friend...that, you, Carol, certainly can control. What you need to be able to do is to release the Karma. Release the bond, so you can forgive yourself fully. Forgive yourself for having done things that were not for the highest good. You need to be able to release that to let yourself off the hook. Can you do that today?"

"Yes."

"All right. What you need to do is say this. I forgive myself. I release any Karma from the past, present, or future, from this life, from all lives, for I have learned my lesson. And I shall not repeat it again. For I have truly learned the error of my ways. I have absorbed this lesson on every level of my mind, body, and spirit. Spiritual support group, has she truly forgiven herself? Look within yourself, Carol. Listen to what they say. Be honest."

"Yes."

"Now, let's turn our attention to Danny Malone. Are you ready to forgive him, Carol? Are you ready to forgive him?"

"He needs to change."

"What does he need to change?"

"His control."

"Okay. But if you are still angry with him, as long as there are any ties that bind, Carol, that control is going to continue. Even perceived control will continue. Forgiveness, grace, wisdom, and love overcome Karma. If you can release him with love, the control is gone. If you're having trouble with this, Carol, I understand. Ask the spiritual support group to help you become willing to forgive. Ask them to help you. Release the ties that bind once and for all. Are you ready to forgive him?"

"Yes."

"All right. Now spiritual support group, we call upon Danny Malone to appear before her right now. Masters and teachers, Archangel Michael, stand by and protect. Now, we call one Danny Malone. We call him up to the platform. He stands in the presence of the spiritual guides and masters. You, too, now call out to Danny Malone. You will hear. Is he there? Is he there now, Carol?"

"Yes."

"Now, Carol, say to him: I forgive you for all that has ever happened in this life or any other life. Our Karma is complete, Danny. There is no tie that binds us. I release you with love, and now it's time to go to the light.

"Tell me what he says or does. Trust your impressions. What's happening, Carol?"

"He's resisting."

"He's resisting. Ask him now. You're safe and protected. Your spiritual support group and guides are around you. Why is he resisting?"

"Control."

"Okay. Ask him: do you love me, Danny? Do you love me, Danny? What does he say?"

"He's crying."

"Okay. Now I'm going to say this, and he will hear it. It's okay if you cry, Carol, because it's good to release. It's good to release."

In my altered state, I could feel the tears gliding down my cheeks. It was a release from Danny's attempt to control me. But the feeling of love from the other side will always be there.

"I'm going to get a tissue and gently wipe your face. He is going to hear this. Danny Malone, you know and I know control is not love. Love is unconditional. Love releases and lets free. If you truly love this woman, you will release her. Right now, you will return to the light and know that you will be loved. Return to the light. As you enter the light, you're surrounded by unconditional love.

"Spiritual support team, we ask that you bring spiritual energy here and now. Bless this one who knows Danny Malone. Raise his vibration. Show him the path to happiness on the other side. Release the flesh and let him go about his spiritual business.

Release him so that he may grow spiritually, releasing Karma, releasing Dharma, expressing unconditional love and all those energies right now. Carol, what is happening now?"

"He's leaving."

"Is he going toward the light?"

"Yes."

"Send the Emissary's spiritual support team to assist him. Make sure he goes to the light. Spiritual support team, make sure this soul truly goes to the light. Go with him all the way. Go with him. Take him to the light. Watch and let me know if he truly goes into the light, Carol. Do you still perceive him?"

"He's gone."

"Good. Spiritual support group, I want you to heal her heart. Heal her emotionally. Heal her on every level. Fill every atom of her being with pure healing energy. Let this be your reality, Carol. Pure white light is filling every area of you. Your heart especially is being healed. Pure love, divine energy. Bask in this. Feel this. Let it fill your entire being, Carol. Feel that healing energy. Let the spiritual support group --your guides, teachers and masters -- heal you and bless you. The energy will continue to fill you long after you have left this room. Your guides and masters will continue to protect you in a very positive and beneficial way. We're not finished yet. I have at least one or two more questions to ask you for the spiritual support group. Right now, I want to ask the spiritual support team: What does Carol need to do to realize her life's true purpose? Listen to what they say, Carol. Tell me what they say or show you."

"To write."

"Good, good. Is there anything that blocks her from this path? Relax and tell me what they say. Trust your impressions."

"Fear."

"What is she afraid of? Be honest, Carol. Look within yourself."

"Rejection."

"I understand. I understand. Okay, I'm going to say this to

you now. Fear is a bitter poison, a false reality. Damage caused by fearing can be real. Do you believe that your self-worth is tied to your writing? You need to release some creative energy. Follow the flow. Let it happen. People will always criticize. Some people will always reject. Some people never accept. That's what is. You can't control other people, but you can control how you respond to them. Let yourself be free. Let yourself express yourself in the way that is best for you. Spiritual support group, is there any particular thing she can do to help her overcome her fear?'

"Focus."

"What does she need to focus on?"

"Experience."

"Okay, spiritual support team, can you help her focus? Can you help her release her fear? What are they saying?"

"They gave me a story."

"They gave you a story."

"Yes."

"You will have total recall of this story. You will remember details of this story. Ask the spiritual support team to help. What is it you need?"

"Insight."

"Okay. What is it you would like to have insight into?"

"Feelings and emotions."

"Okay. This is a story that illustrates something for you. Is it a story you're supposed to write?"

"Danny."

"Okay. I'm going to give you a suggestion right now. You will clearly remember this story. You will have total memory of this story upon awakening. You will be full of motivation. Whether writing it on a tablet or a computer, you will have total recall of all that has transpired today. Feelings and emotion. Okay, spiritual support group, help her right now. Please share with me what you perceive, Carol."

"Polarity."

"Okay. Do you know what this phrase means? Do you know

what this word means to you? Is it clear to you, Carol?"

"I'm not sure."

"All right then. Spiritual support group, can you supply more insight into this for her. Give her understanding of it?"

Time passes -------to collect my thoughts. ***

"He has the gift."

"Yes. So do you, Carol. So do you, Carol. You have the gift. You grow stronger and stronger every single day. You're a gifted writer. Now in your body, where is the blockage?"

"In my torso. Center."

"What color is it? Give it a color."

"Brown."

"And if it had a shape, what shape would it be, Carol?"

"Like a block."

"Okay, go ahead and take this shape out with your left hand. Physically do this, please. With your left hand, take this shape out of your torso and throw it away. Get rid of it. Give it back to wherever it came from. You can do it. Throw it out and give it back to whoever gave it to you. You know who that is. You can just simply toss it out. Is it all gone?"

"Yes."

"Okay. Now, in your right hand I want you to imagine that you have the powers of expression, confidence, serenity, well-being. Confidence, serenity, well-being. Confidence in your expression, the ability to write in free expression. Well-being, freedom of expression. Give that a color. What color would that be, Carol?"

"Red."

"Okay. What kind of shape would it be?"

"Fragmented."

"If it's fragmented, pull it together. Give it a shape that means something good to you. Let me know when you have that."

"Heart."

"Okay. Go ahead and take that shape, that red shape, a heart. Go ahead and put it where that yucky brown thing was. Physically do it with your right hand when you are ready. Go ahead. Put that energy there. That's right. That's right. Just feel that energy filling your body. Cleaning your torso, expanding and filling your entire being. That beautiful red light is filling you completely, Carol, from head to toe. In a moment, we are going to get ready to emerge you. But before we do that, I'm going to ask the spiritual support group: is there anything else they need to say to you today? Any words of wisdom or advice?"

"Trust."

"That's right. Trust yourself, Carol. Trust yourself. Trust your guides and masters. Your spiritual growth is assured. Your spiritual growth is a certainty. All right. I'm going to go ahead and emerge you from the altered state. I'm going to do that by counting from one to five. At five, you will be able to open your eyes. Notice how good you'll feel. You'll feel good the rest of the day. You'll have a clear mind. You'll be able to drive yourself home. You'll be able to write down that story which you are going to remember. One, two, three, four, five. Open your eyes, Carol and notice how good you feel."

"Wow, I don't want to move right now. I'm so relaxed. Thank you. Thank you for all your help. I have been struggling with this for so long. It's has been difficult to remove Danny's energy. He has been with me for several years now."

"I understand, Carol. I understand."

I wiped my tears from having a deep emotional experience and removed myself from Bob's comfortable recliner. Then I shared a hug with Bob and gathered my purse to leave. "Thank you again."

"If you know anybody who needs help, please give me a referral."

"I certainly will. You're the best, Bob."

While driving home I had to ask myself, *Is Danny really gone?*

The following evening, I recited a meditation prayer to the Angel Raphael. I wanted to bring peace and light to Danny and

Alex. Whatever the conflict or dispute that might have been created in their energy field, I knew they could not find healing until they were free from their earthly egos.

I lit a candle on my glass table and began, "Raphael, angel of happy meetings, lead us by the hand toward those we are looking for. May all our movements and all their movements be guided by your light and transfigured by your joy.

"Angel guide of Tobias, lay the request we now address to you at the Feet of Him on Whose unveiled Face you are privileged to gaze.

"I, Carol Shimp, ask for love, light, and peace, for Alex Shimp and Danny Malone.

"Lonely and tired, crushed by the separations and sorrows of earth, we feel the need of calling to you and of pleading for the protection of your wings, so that we may not be strangers in the province of joy, and ignorant of the concerns of our country.

"Remember the weak, you who are strong -- you whose home lies beyond the region of thunder, in a land that is always peaceful, always serene, and bright with the resplendent glory of God. Amen."

CHAPTER 31

Love

It was the tenth of May, 1998, Mother's Day. I hadn't planned to move on Mother's Day, but I just happened to have the day off from work. I had on my usual blue jeans and blue T-shirt. Tom had emptied, cleaned, and painted his bedroom closet. Tom just lived across the street from the grocery store. I could still walk to work. I had agreed to give my son, Jerry, my furniture. Packing and stacking boxes in my apartment, I knew Princess would like her new home. She could lie on the floor by the big front window and gaze outside.

I received a message in my thoughts. *"This isn't over yet."*

Princess had her nose to the front window and yelped as Tom backed his truck up to my apartment door. He knocked and I opened the door. He stood there in white T-shirt and jeans. He liked white T-shirts. We gave each other a hug. I gazed into Tom's blue eyes. Again, I thought I saw a familiar light there, in his eyes. He stooped down to greet and pet Princess.

"You're going to like your new home, girl. We are going to have so much fun."

Tom stood up and we each lifted a box to load in the truck.

Then, a song came on the radio in my apartment. I was

shocked to hear, *"I want to know what love is. I know you can show me."*

"That's weird, Carol. Is the radio plugged in?"

"No. I unplugged it so we could pack it in the bed of the truck."

Tom shrugged his shoulders.

My mother had said through me in automatic handwriting just before Alex died,

"So many times we come by the way of love only to lose it."

I thought of Alex and Danny, the Karma we created. Action promotes reaction. Lovers come and go. Years drift away. In this circle of life, we all try to hold on to Spirit, to command and faithfully commit to lasting relationships. It's that Karma word that bothers me.

When we arrived at Tom's home, Princess jumped out of the truck. We stepped up to the front porch and Princess sniffed the door; then she looked at Tom and barked twice to give her approval. I was surprised because the whole family greeted me. What a wonderful Mother's Day. Janet and Jerry were there. Tom's son Jerry, his daughter Marie, and her son Kent were there.

June walked toward me and we shared a hug. "Welcome to the family. We ordered pizza and we have cake for dessert."

"That's great -- we won't have to worry about cooking dinner." I looked at the Mother's Day cake that sat in the center of the dining table. It looked yummy, with pink roses and creamy icing. Everyone helped carry boxes in the house. My Jerry gave me pink roses and Janet handed me a package to unwrap. I was pleased to find pink slippers, and then the three of us shared a hug.

Jerry, my son, scraped his finger across the side of our cake and placed icing on Princess's nose. Everyone laughed.

Tom was extremely joyful, smiling and humming. I could tell he liked having his home filled with family and friends.

I wanted to cry, but held back the tears. Family had always meant so much to me. Now with my children and Tom's relatives together, I felt blessed. My mother's spirit had walked with me, through my troubles and heartache. She brought me home again to a new family.

I couldn't believe Tom had planned this surprise for me. I stood alone for a moment on Tom's front porch. I felt spiritual energy. Then, I heard in thought, *"My family."*

Tom's mother, in spirit. "Betty," I said. Red had called me Betty long before Alex died. Betty's spirit was leading me to Tom. She wanted happiness for her son. Our mothers were working together on the other side. *Wow! My testimonial from Betty and my mother to love from the other side. What a wonderful magical moment this was!*

Tom opened the front door and we hugged each other. "I love you from this side," he said, reached for my hand, and guided me inside.

Alex's spirit visits me once in a while at Tom's house. Princess confirms his presence. She will whine or cry that special sorrowful sound of hers. I know Alex's spirit comes just to check and see if we are okay.

THE END

Afterword

Originally, I had thought to seek answers from the Catholic church. But in reality, being raised a Catholic I knew it would be a long political process. The priest would have to get permission for an exorcism from the cardinal and eventually all the way to the Pope. I wonder if, when a priest gives a confessor absolution, does that dissolve our Karma?

In the beginning, as Danny's spirit became more dominant, I decided I did not want an exorcism. I wanted answers. I wanted to know why Danny's spirit wouldn't leave and go into the light.

Eventually, as I was finally freed, I was grateful for Red's help. He knew I needed to be free from Danny's energy. I was never terrified of spiritual energy. My experience has taught me that earthbound spirits are more intense, while our guides are gentle and have a brighter light.

When I was in a deep sleep and traveled into God's light with spirit and learned my prophecy, I was surrounded by love, and was told I would be protected. I couldn't remember the entire message I received. As events happened I knew intuitively that I was going to learn more. Danny and I shared a Karmic relationship.

Danny, in my automatic handwriting, wrote, *"Find the truth."*

I believe the truth is that love never dies. In my regular meditation I pray that Danny is where he is supposed to be and that he has found peace.

My memories were always there to trigger my emotions. We retreat back to that first meeting, when we gravitated to the attraction of love and mystery.

I'm grateful for the wonderful experience my spirit guides have given me. I was blessed with psychics who understood my dilemma and were able to help me. The message in one of my dreams was, "The key is love."

The answer is: God's love heals all.

Now I enjoy my evening walks in Tom's neighborhood. The streets are wider, and the homes are groomed with greener lawns and flowers.

The statue I had given my parents of the old couple dancing for their 50th wedding anniversary now has a place in Tom's china cabinet. And yes, he still helps with the laundry.

Acknowledgements

John Bandy, Thank you for your art work and my book cover. You are truly a blessing.

To Margaret, my mother. Who taught me, "Love is the key." Her spirit walked beside me through my grief, heartache, and joy.

To the memories of my husband Albert, and my son Jeffrey. Their loving light has helped me find comfort and peace.

To my present companion, and my daughter. They have stood by me and put up with all of my difficult times and doubts.

To my son-in-law, for his computer genius. To quote my publisher, "He is a gift from God." I'm grateful for your time and dedication.

To the earthbound spirit, who opened the door and taught me to accept my spiritual gift, as I helped him on his journey to receive love, peace, and healing.

To Marjory D. Lyons, and Susan Glazer, for helping me with missing parts and rewrites. Thank you for understanding my quest to tell my story, and also for your dedication and abundance of knowledge.

I want to thank my writer's circle, for listening to my rewrites and their patience to sit through my blunders. Eleanor Rostad, Sharon Wharton, Marge Bowman, Ruthie Mesch, Lois Rudnick, and all those who stood by me.

To Mark Dodich, my nephew, thank you for your astrological and spiritual counseling. Your honesty and trust are cherished. www.astromark.us

Thank you, Reverend Edward "Red" Duke, for your home, The Haven For Spiritual Travelers, and for the healing power of Spirit. Thank you for helping me through my grief and helping me discover answers to the mystery of my earthbound entity. It is my desire that through my experiences of joy, grief, and suffering, others will be touched by the healing power of Spirit. My testimonial to Red: five years ago I had a lumpectomy for breast cancer. During my last mammogram, I was told they did not even find scar tissue. www.reverendredduke.com

Thank you Loretta Rohr, for your gift of channeling and friendship.

To Bob Decker, thank you for my regression and opening the door to peace. Thanks for making it easier to reveal the truth, that spirits need healing as much as the living, despite those who use unawareness and fear to deny it. www.bobdecker.net

To Anne Miller, thank you for your intuitive reading and insight.

I also want to thank Pepper May. You were kind enough to open your home for meditation and creating a beautiful white light. During the 1980s, I enjoyed meeting friends and practicing hands on healing in Pepper's home.

Phil and Linda, of "Crystal Fantasy," thank you for your Reiki meditation circle on Friday evenings. It has been comforting and healing. www.crystalfantasyfla.com

Books That Have Helped Me

The Donning International Encyclopedic Psychic Dictionary by June Bletzer, Ph.D.

Reaching to Heaven by James Van Praagh

What If God Were the Sun by John Edward

The Celestine Prophecy, The Tenth Insight by James Redfield.

Life and Teaching of the Masters of the Far East by Baird T. Spalding

Healing Lost Souls, Releasing Unwanted Spirits From Your Energy Body by William J. Baldwin, Ph.D.

I am at Bell Rock in Sedona, AZ. The sun is reflecting the energy centers on my right hand.